Metal-Organic Frameworks as Forensic Detectors

Vivek Mishra • Amarnath Mishra

Metal-Organic Frameworks as Forensic Detectors

Springer

Vivek Mishra
Amity Institute of Click Chemistry
Research and Studies
Amity University
Noida, Uttar Pradesh, India

Amarnath Mishra
Amity Institute of Forensic Sciences
Amity University
Noida, Uttar Pradesh, India

ISBN 978-981-96-2146-0 ISBN 978-981-96-2147-7 (eBook)
https://doi.org/10.1007/978-981-96-2147-7

This Springer imprint is published by the registered company Springer Nature Singapore Pte Ltd.
The registered company address is: 152 Beach Road, #21-01/04 Gateway East, Singapore 189721, Singapore

Dedicated to our beloved parents

Preface

In the intricate realm of forensic science, the quest for precision and reliability in detecting trace elements has led to the exploration of cutting-edge materials. This book, *Metal-Organic Frameworks as Forensic Detectors*, embarks on a journey through the fascinating intersection of MOFs and forensic investigations. MOFs, characterized by their exceptional porosity and tunable properties, emerge as pivotal tools in unraveling the mysteries concealed within forensic samples.

Chapter 1 introduces the readers to the fundamental aspects of MOFs, providing a comprehensive understanding of their structural intricacies and diverse applications. This lays the groundwork for a profound exploration of MOFs in forensic science, where their unique characteristics are harnessed to address a myriad of challenges.

Chapters 2 and 3 delve into the specific realms of photochemical and electrochemical properties of MOFs, unraveling their potential in augmenting forensic analysis techniques. The subsequent Chaps. 4, 5, 6, 7, 8, 9, 10, 11 and 12, unfold a captivating narrative, each dedicated to a distinct forensic application of MOFs. From detecting latent fingerprints and lip prints to identifying pesticides, antibiotics, drugs of abuse, and hazardous chemicals, MOFs emerge as versatile agents in the forensic scientist's toolkit.

In Chap. 13, the book critically examines the advantages and disadvantages of employing MOFs in forensic sample detection. This insightful analysis serves as a guide for forensic practitioners, offering a nuanced understanding of the practical implications and limitations associated with MOFs.

The book navigates through the current challenges faced in the integration of MOFs in forensic science, paving the way for a comprehensive discussion on prospects. As we stand at the forefront of technological advancements, the potential for innovation and refinement in MOF applications within forensic science is limitless.

This book is not merely a compilation of scientific endeavors; it is an exploration of possibilities. It beckons undergraduates, post-graduates, researchers, forensic scientists, and enthusiasts alike to embark on a journey that transcends the conventional boundaries of forensic science. In the synergy between MOFs and forensic investigations lies the promise of unraveling the most intricate puzzles concealed within forensic samples.

Noida, Uttar Pradesh, India

Vivek Mishra
Amarnath Mishra

Contents

About the Authors

Vivek Mishra currently serves as an Associate Professor at the Amity Institute of Click Chemistry Research and Studies, Amity University, Uttar Pradesh, Noida, India. He earned his Ph.D. in Chemistry from Banaras Hindu University and honed his skills through postdoctoral positions at prestigious institutions in South Korea as well as in India. Dr. Mishra's research interests encompass a diverse range, including forensic sciences, organic chemistry, and advanced functional materials. With over a decade of teaching experience across various chemistry disciplines and forensic sciences, he has made significant contributions, including 4 books, 20 book chapters, 60+ research articles, and 10+ international and national patents evident in his extensive publication records. Additionally, Dr. Mishra actively participates in scientific societies, such as the American Chemical Society, National Environmental Science Academy, and Worldwide Association of Women Forensic Experts, further enriching the academic community with his insights and collaborations.

Amarnath Mishra is currently an Associate Professor at Amity Institute of Forensic Sciences, Amity University Uttar Pradesh, previously led as Founder Head at Lloyd Institute of Forensic Science. With over 16 years of expertise, he earned his Ph.D. in Forensic Science specializing in Toxicology, M.Phil. in Biochemistry, and M.Sc. in Forensic Science. Dr. Mishra's contributions include 7 books, 25 book chapters, 60+ research articles, and 20+ invited talks globally. He supervised Ph.D. students, guided 150+ undergraduates and postgraduates, and trained over 100 Investigating Officers. Acknowledged with National Academic Excellence and International Distinguished Academic Leader awards, he serves as a Freelance Forensic Expert and Medico-legal Consultant while actively contributing to professional bodies.

Introduction of Metal-Organic Frameworks (MOFs)

1

Abstract

The term "metal organic frameworks" (MOFs) refers to a class of compounds, either metal ions or clusters, that are balanced with organic ligands to form one or more dimensional structures. Reticular synthesis generates MOFs, whose precise constituent selection can yield crystals with ultrahigh porosity and strong bonds between inorganic and organic units. It has not been demonstrated that other solids may increase their metrics without changing the underlying topology, nor can they demonstrate the same accuracy typically employed in chemical modification. MOFs may yield compounds that offer a synergistic combination of properties due to the form of building units and chemical compositions multiplying based on shapes. This chapter covers recent developments in MOF synthesis techniques as well as structural features. Additionally, it provides a brief overview of the various forensic fields in which MOFs are utilized.

1.1 Introduction: A Brief History and Key Facts

In 1965, Tomic's first article about porous materials and metal organic frameworks (MOFs) piqued the interest of scientists at the time. The significance of materials is decided by the function they play in everyday life and their impact on the future of human technology. MOFs are a type of porous inorganic-organic hybrid component made up of metal oxo clusters linked by organic ligands [1]. MOFs have both organic connections and metal centers in their structure. Organic links are considered organic SBUs (Secondary Building Units) that operate as "struts," whereas metal centers are inorganic SBUs that act as "joints" in MOF formations [1]. The architecture of the framework, inorganic metal centers, and organic ligands are the three primary components of MOF. If the organic and inorganic structures of MOFs were created using enlarged zeolite topology [2], they would have more pores and

V. Mishra, A. Mishra, *Metal-Organic Frameworks as Forensic Detectors*,
https://doi.org/10.1007/978-981-96-2147-7_1

higher porosity than zeolites. The network's dimensions caused by large metal clusters and the enlarged size of inorganic SBUs are the causes of increased pore size. Moreover, extended organic ligands that benefit from a wide pore diameter and high porosity are responsible for the length of link between inorganic SBU. As molecular building blocks, MOFs are created by applying various metal centers and ligands, which leads to an appropriate flexibility for changing chemical and physical properties [3–6]. MOFs are known to provide a superior platform for immobilizing molecular catalysts on conductive substrates because they have the largest pore volume and specific surface area. These materials can also be modified to have a high selectivity for a certain chemical process. According to theoretical estimations, MOF semiconductors' optical band gap can be set between 1.0 and 5.5 eV [7–9]. The self-assembly polymerization of solid interatomic linkages is not restricted to a single plane. It is carried out by metal ions or clusters and stiff organic molecules connected by coordinative bonds. This polymerization produces incredibly crystal-like constituents that are significantly dragged out into three-dimensional structures. MOF emergence is well acknowledged as a method to mimic inorganic substances, such as zeolites. Actually, what was achieved in the design and synthesis of hybrid (extended) constituents with stable microporosity was generalized in this way [10–14]. In contrast to MOFs, zeolites do not permit precise control over the functionalization, size, and shape of pores. The best and optimal constituents are those that are easily derived, consistently applied, and require little effort. Theoretically, MOFs can fulfil these demands in a straightforward way because synthetic design principles are fundamental and only require careful primary selection of metal centers and ligands. In the solid state, they can self-assemble under certain synthetic conditions. The endless world of hybrid organic-inorganic combinations can be guaranteed by the great diversity of such basic components. In an effort to forecast their architectures, the literature study did reveal the existence of several typical structural motifs in the creation of MOFs [15–19].

Organic molecules are typically represented by molecules containing one or more donor atoms, either N or O. These molecules operate as a kind of link between metal ions. Among the common ligands are polyamines (more especially, those originating from imidazole, oxalic acid, and benzene), carboxylates, pyridyl and cyano-groups, crown ethers, and phosphonates. This method of molecular manipulation was used by chemists to predict network topologies and control their shape, size, and dimensionality while also verifying the problem of MOFs' chemical modification. Further, by adding functionality to the linkers and applying a particular reactive group, chiral center, or redox center, a targeted feature can be achieved across the bulk material. The development of new linkers and the substitution of commercially existing compounds can be greatly impacted by conventional organic synthesis or, alternatively, in situ production techniques. On the other hand, there is a relationship between the coordination geometry and the end product's dimensionality: the framework's topology is influenced by the electrical configuration, coordination modes, and the size or hardness of the metal center [20–25]. This book aims to provide a quick overview of the improved MOF-based method's performance was contrasted with existing, widely used methodologies.

1.2 Metal Organic Frameworks (MOFs): Classification and Nomenclature

The term "MOFs" stands for metal organic frameworks; it is typically used as a collective term for the class of compounds. When it is followed by an ordinal number, it refers to a specific metal organic framework. The examination of MOFs' characteristics and configurations may aid in the creation of framework structures with the intended attributes, such the IRMOF-1–IRMOF-16 family of MOFs (isorecticular metal–organic frameworks) that share the same symmetry [26, 27]. Numerous MOFs are assigned to organizations (materials of institute Lavoisier), including UiO, MIL, HKUST, LIC, etc., according to where they were discovered. Chinese [28, 29] and Russian [30–32] researchers frequently use the term "metalorganic coordination polymers" with defined composition (Fig. 1.1). The other big family of MOFs is zeolite topology. Imidazole rings, which have varying functions, join metal ions like Fe, Co, Cu, Zn, and so forth, and encircle them with tetrahedra composed of nitrogen atoms. These MOFs can be created with the ZIF (zeolite imidazolate framework) abbreviation and a number. The study groups' classification of synthesized MOFs, such as CPL-2, F-MOF-1, and MOP-1, is derived from numerous different classifications [33] (Table 1.1, Fig. 1.1).

Fig. 1.1 Representation of MOFs structures. (Reprint with permission from Ref. [33]. Copyright @2019 Elsevier)

Table 1.1 A variety of common MOF designations and their compositions

S.No.	Abbreviation interpretation	Designation	Formula
1	Metal-organic frameworks	MOFs-74	Zn_2DOT
		MOF-177	$Zn_4O(BTB)_2$
		MOF-101	$Cu_2(BDC\text{-}Br)_2(H_2O)_2$
		MOF-253	Al(OH)(BPYDC)
		MOF-235	$[(Fe_3O(BDC)_3(DMF)_3][FeCl_4].(DMF)_3$
2	Isoreticular metal-organic frameworks	IRMOF-1 (MOF-5)	$Zn_4O(BDC)_3.7DEF.3H_2O$
		IRMOF-16	$Zn_4O(TPDC)_3.17DEF.2H_2O$
3	Materials of institute Lavoisier	UiO-66	$Zr_6O_6(BDC)_6$
		UiO-67	$Zr_6O_6(BPDC)_6$
		MIL-53	Al(OH)(BDC)
		MIL-88A	$Fe_3O(MeOH)_3(O_2CCH{=}CHCO_2)_3.MeCO_2.\ nH_2O$
		MIL-100-Fe	$Fe^{III}{}_3O(H_2O)_2F.(BTC)_2.\ nH_2O$
		HKUST-1 (MOF-199)	$Cu_3(BTC)_2$
		LIC-1	$Gd_2(BDC\text{-}NH_2)_3(DMF)_4$
4	Zeolite Imidazolate frameworks	ZIF-8	$Zn(MIM)_2$
		ZIF-90	$Zn(FIM)_2$
5	Coordination polymer with pillared layer structure	CPL-2	$Cu_2(PZDC)_2(4,4'\text{-}BPY)$
6	Fluorinated metal-organic frameworks	F-MOF-1	$[Cu(HFBBA)(phen)_2](H_2HFBBA)_2(H_2O)(HCO_2)$
7	Metal-organic polyhedra	MOP-1	$Cu_{24}(m\text{-}BDC)_{24}(DMF)_{14}(H_2O)_{10}$

Note. DOT is 2,5-dihydroxyterephthalate, BTB is benzene-1,3,5-tribenzoate, BPYDC is 2,20 -bipyridine-5,50 -dicarboxylate, TPDC is p-terphenyl-4,40 -dicarboxylate, BPDC is biphenyl-4, 40 -dicarboxylate, MIM is 2-methylimidazolate, FIM is 2-formylimidazolate, PZDC is pyrazine-2,3-dicarboxylate, 4,4′-BPY is 4,4′-bipyridine, phen is 1,10- phenanthroline, HFBBA is 4,4-(hexafluoroisopropylidene)dibenzoate, m-BDC is m-benzenedicarboxylate

1.3 Synthesis Techniques of MOFs

MOFs are now acknowledged as a significant class of porous compounds due to their unique functional and structural characteristics. Metal ions, metal ion clusters, and organic linkers that bridge are what form these frameworks. The chosen main building blocks (PBUs) are without a doubt another idea that plays a big role in the final structure and attributes of the MOF. However, there are numerous additional synthetic techniques and parameters that need to be considered as well, like temperature, reaction time, pressure, pH, and solvent. Depending on the final structures and characteristics, a variety of synthetic techniques, such as slow diffusion [34–37], hydrothermal (solvothermal) [38–41], electrochemical [42–44], mechanochemical [45–48], microwave assisted [49–52] heating, and ultrasound [53–56], can be used to create MOFs.

1.3.1 Slow Diffusion Technique

The conventional diffusion method involves introducing different species into interaction one at a time: (i) Soluble liquid diffusion is one approach. First, we create two layers, one containing the precipitant solvent and the other encasing the product in a solvent, both of which have different densities. Remarkably, there is a solvent layer between them. The progressive diffusion of the precipitant solvent into the distinct layer is followed by crystal formation at the interface; (ii) The reactants are progressively dispersed in the second technique by physically dividing barriers, such as two vials of different capacities. Furthermore, gels are used as crystallization and diffusion mediums in some circumstances, specifically to slow diffusion and inhibit bulk material precipitation. The diffusion process is predicted to produce single crystals suitable for X-ray diffraction analysis as a substitute for non- or polycrystalline products, especially if the materials are not very soluble [34, 35, 37, 57].

1.3.2 Hydro (Solvo) Thermal Technique

The process of hydro (solvo) thermal self-assembly of products from soluble precursors is employed. Although this problem was initially used in the synthesis of zeolites, it has now also been applied to the synthesis of MOFs. The operating temperature range in an autoclave under autogenous pressure is between 80 and 260 °C, and it is mostly dependent on the rate of cooling once the reaction is finished. However, a number of situations necessitate lengthy response times—several days in the case of solvothermal and hydrothermal procedures and many weeks in the case of the diffusion approach [38–41, 58].

1.3.3 Microwave Technique

Microwave-assisted technique results in the creation of small metal and oxide particles. For the purpose of producing metal's nanosized crystals, such techniques may involve heating the solution with microwaves for an hour or longer. Their application in the production of crystalline MOFs has not been very frequent thus far. Nevertheless, they are an essential technique for quickly carrying out synthesis. Furthermore, this approach allows for proper control over the size and shape of the resultant particles. Even while the microwave technique usually cannot generate crystals, it has an impact comparable to single X-ray analysis in that it can produce a consistent seeding condition. As a result, we can adjust the size and shape of the crystals and significantly reduce the synthesis cycle for several operational uses [49–51, 59]. The method of solvent evaporation requires certain conditions, such as the formation of crystals in saturated solutions and temperature-dependent solubility improvement, which causes crystals to form during the cooling phase [60].

1.3.4 Electrochemical Technique

MOF powders are produced using an electrochemical process on an industrial scale. Compared to solvothermal synthesis, this approach has several advantages, such as avoiding anions like nitrates from metal salts, milder reaction temperatures, and extraordinarily fast synthesis. The generation of metal ions in situ close to the support surface limits the amount of crystallization that can occur in the bulk step and is thought to reduce the unfavorable accumulation of crystals during membrane construction. In addition, it appears that the lower temperatures are helpful in this regard since thermally induced cracking during the cooling phase (as opposed to solvothermal synthesis) must be less distinct. This breaking is caused by an incompatibility in the thermal growth coefficients between the various support structure and the MOF. MOF exhibits a negative thermal expansion coefficient in this regard [42–44]. Because of the straightforward voltage adjustment or imposition of specific signals (e.g., pulses), electrochemical synthesis methods yield more parameters for finetuning than solvothermal procedures.

1.3.5 Mechanochemical Method

Mechanical force can produce a variety of physical events, including chemical reactions and mechatronic processes. In mechanochemical synthesis, the mechanical breakdown of intramolecular bonds occurs before a chemical transformation [61, 62]. Mechanochemistry has a long history in synthetic chemistry and is currently used in many areas such as inorganic solid-state chemistry, organic synthesis, polymer science, and multicomponent (ternary and higher) reactions to create pharmacologically active co-crystals. There are several good reasons to be interested in mechanically activated MOF synthesis. Environmental concerns are therefore among the important points. Reactions can be carried out at room temperature in the absence of organic solvents, which is particularly helpful in situations where it is not possible to employ organic solvents. It is possible to produce products with tiny component counts and to achieve quantitative yields in 10–16 min reaction times [63, 64]. Additionally, the only byproduct produced by metal oxides as a starting material and occasionally utilized in place of metal salts is water. However, due to their poor solubility, metal oxides are rarely used in solvent-based processes. Applying hydrated metal salts and organic reactants with low melting points, particularly those containing important anions, appears to be beneficial. In order to obtain well-crystalline compounds, metal acetates or carbonates were employed. Thus, although acetic acid is known to be a byproduct found in pores, it is easily removed via heat activation [65, 66]. Liquid-assisted grinding (LAG), which adds minuscule amounts of solvents, has been shown to offer significant advantages. This procedure quickens the mechanochemical processes, maybe because of increased molecule-level reactant mobility. The liquid might simultaneously reveal features that direct the construction. As of right now, it has been established that the technique may be expanded to ion- and liquid-assisted grinding (ILAG) for the selective manufacturing of pillared-layered MOFs with great efficiency, proving that ions and solvents have a structure-directing effect [67–69].

1.3.6 Sonochemistry Technique

Sonochemistry is the study of chemistry that takes place when a reaction mixture is exposed to high-energy ultrasound [70]. Ultrasound is cyclic mechanical vibration at 10 MHz, the highest limit of human hearing, and a frequency between 20 kHz and higher. The production of microjets during cavitation is accompanied by the ability to corrode, activate, and clean any nearby solid surfaces. Furthermore, there occurs the agglomeration-dissemination of smaller particles. In a homogeneous liquid, chemical reactions can take place in the bulk media, where there are significant shear forces, at the interface (intermediate pressures and temperatures), or in the cavity (severe conditions). Severe circumstances and significant shear forces have the potential to create radicals with an extra reaction, break bonds, and put molecules in an excited mode (see mechanochemistry). Moreover, ultrasonication can improve the initial compounds' dissolution concurrently. Sonochemistry is widely used to carry out the synthesis of organic and nanomaterials. The primary goal of MOF science's sonochemical synthesis process is to create a fast, energy-efficient, environmentally benign, user-friendly, and ambient temperature application technique. The fact that MOFs scale rapidly through fast reactions has piqued curiosity in their potential applications [71, 72].

Numerous techniques, combined with the abundance of readily available components and the variable process conditions, have developed thousands of MOF compounds.

1.4 MOFs: A Versatile Platform for Forensic Detections

Many of the current methodologies that are regularly utilized in forensic laboratories around the world have been modified from other disciplines. Similarly, the guided growth of Metalorganic Frameworks (MOFs) was the subject of a recent paper in the field of material sciences. MOFs have a large interior surface area and a very porous crystal structure. Due to these qualities, they are particularly helpful in a variety of applications in the forensic industry.

1.4.1 Photochemical and Electrochemical Properties of MOFs

As discussed earlier, MOFs are a class of porous crystalline materials possessing distinct optical, electrical, and catalytic capabilities along with well-organized structures (Fig. 1.2). MOFs and materials derived from MOFs have been shown to be highly effective catalysts for water splitting through both electrochemical and photoelectrochemical (PEC) pathways. In addition, by customizing the organic

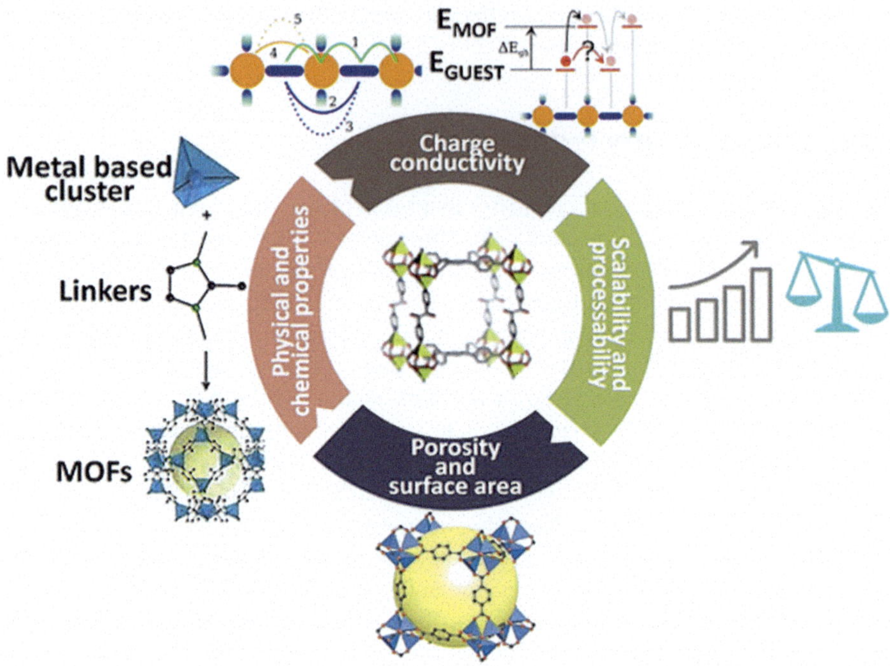

Fig. 1.2 An overview of MOFs based photochemical and electrochemical properties [73]

ligands and/or metal centers, the bandgap, surface area, current density, electrochemical active surface area, and overpotential of these MOFs, their photochemical and electrochemical capabilities can be optimized to maximize their performance. Despite being in its early stages, this topic is garnering more attention due to encouraging findings that point to a remarkable potential for real-world applications [73].

1.4.2 MOFs for Detection of Latent Fingerprint Detection

MOFs are crystalline porous structures that are currently employed as catalysts, separators in membranes, and sensors. They have lately been suggested for fingerprint detection because of their optical and physicochemical characteristics. Their potential for fingerprint detection was further investigated in this work. The technique's effectiveness was evaluated using both naturally occurring fingerprints and marks that were charged, and protein enriched. Numerous parameters were examined and refined, including precursor concentration, pH, immersion time, and detection techniques [75]. Then, the improved MOF-based method's performance was contrasted with existing, widely used methodologies. The results showed that MOFs can detect fingerprints with high specificity, particularly those rich in proteins, such those left by bodily fluids. Nevertheless, findings were found to be lower than those from currently used detection methods after comparison and evaluation against

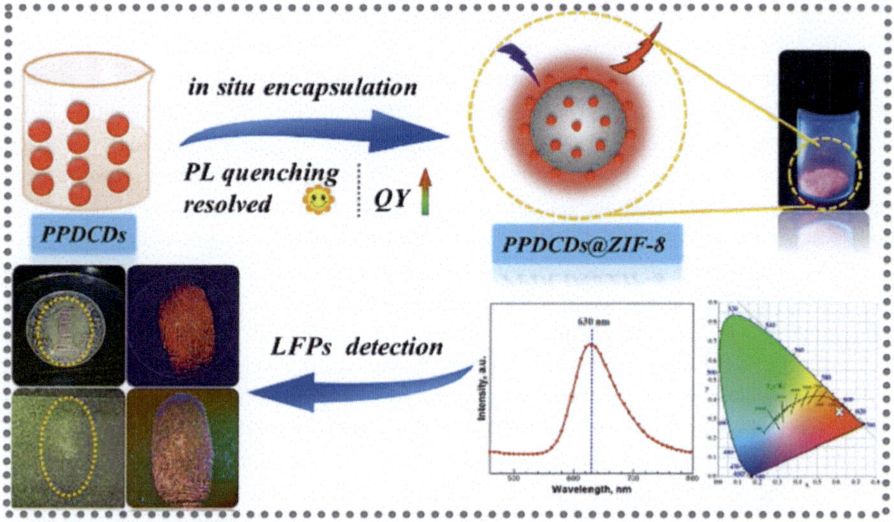

Fig. 1.3 A visual representation of fingerprint detection. (Reprint with permission from Ref. [74]. Copyright 2023 Elsevier)

benchmark approaches. However, MOFs could be a viable substitute for the existing powder suspension methods with more study and improvement. Recently, Zhu group [74] carbonized PPDCD polymer dots were successfully combined in situ with the zeolitic imidazolate framework ZIF-8 to create a highly luminous red emissive nanocomposite material (PPDCDs@ZIF-8). After conjugation with ZIF-8, the resulting material exhibits a strong and steady red emission at solid state, with the largest peak centered at 630 nm. Additionally, the fluorescence quantum yield of PPDCDs greatly increased, rising from 17.8% to 43.5%. The ultrahigh sensitivity of this nanocomposite material has been effectively utilized to detect LFPs on marble, glass, and other surfaces. The acquired fluorescent pictures simply demonstrated the intricate features of LFPs ranging from level 1 to level 3. Because it offers a quick and precise method of seeing the sweat pores in LFPs, PPDCDs@ZIF-8 has a lot of potential uses in crime scenes (Fig. 1.3).

1.4.3 Metal-Organic Frameworks for Detection of Pesticides in Forensic Samples

Global population expansion necessitates the intensification of anthropogenic processes, resulting in hazardous organic compound pollution of both terrestrial and marine habitats. Persistent organic pollutants (POPs) are particularly hazardous manmade chemicals. Due of its comparatively great chemical resistance, POP use is now strictly regulated or banned globally. Unfortunately, certain POPs are still used as agricultural practice aids. The bulk of POPs were once widely used as pesticides. Consequently, there is a growing demand for environmental monitoring to accurately detect and quantify pesticidal POPs. The selectivity and sensitivity of chemical sensors and adsorption materials designed for certain pesticides should enable the detection of target molecules down to the nanomolar range through host-guest interactions. Materials having a very large active surface area, a clearly defined structure, and excellent stability could do this. In that context, innovative materials that are being researched are called metal-organic frameworks (MOFs). Different MOFs can have their structures functionalized to enable desired host-guest interactions. We critically examine the use of MOFs in this book (Chap. 6) for the detection and adsorption of specific pesticides that are categorized as persistent organic pollutants (POPs) by the Stockholm Convention. In this regard, Yan et al. (2020) used the zeolitic imidazolate framework-8 (ZIF-8)/Au-Ag/Si assembly (see Fig. 1.4) as a SERS material in their PCP detection technique. Significant control over the film thickness and, consequently, the SERS effect was exercised in the preparation of this material, which was done layer by layer. It was discovered that the concentration of the target molecule around Au-Ag nanoparticles was caused by the electrostatic interactions between ZIF-8 and PCP, which significantly increased Raman scattering through plasmonic effects. ZIF-8/Au-Ag/Si assembly application for SERS-based PCP detection at sub-pM level [76].

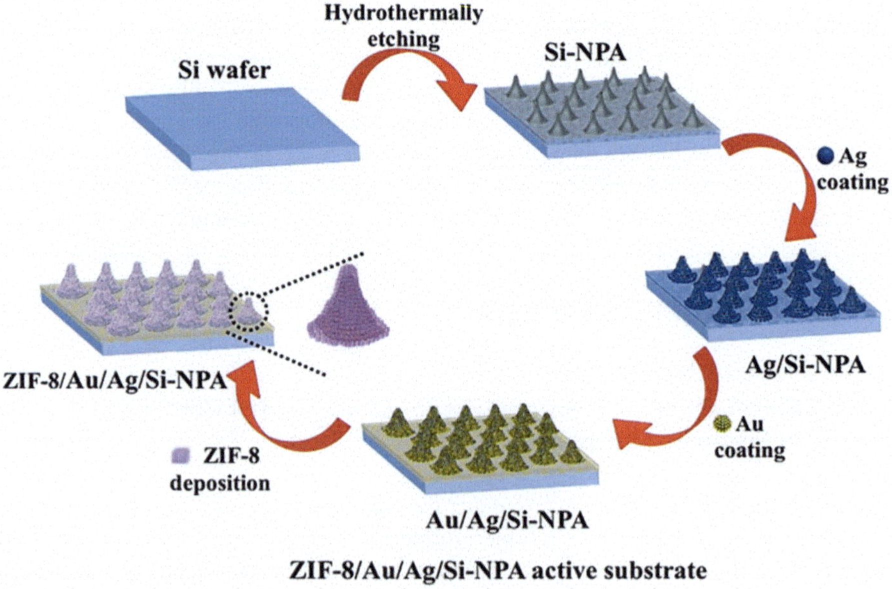

Fig. 1.4 Diagram representation of synthesis of ZIF-8/Au-Ag/Si-NPA active substrate for pesticide detection [76]

1.4.4 Metal-Organic Frameworks for Detection of Antibiotics in Forensic Samples

The significant environmental threat posed by antibiotics has sparked intense research interest in the development of appropriate chemical sensors for the selective detection and quantification of antibiotics in recent years. The creation of appropriate sensors is necessary because the overabundance of antibiotics in the soil, caused by increased human use and animal activity, might have detrimental impacts on living systems. The fluorescence technique has emerged as a viable and efficient alternative to the several detection technologies now in use for the identification of antibiotics. A wide range of fluorescence sensors for the selective detection of antibiotics in solution and vapor phase have been developed recently. Using luminous metal-organic frameworks (LMOFs) as fluorescence sensors for the discriminative detection of various antibiotic kinds, we have made significant advances recently [77, 78]. We endeavored to methodically summarize these developments in this book (Chap. 7). Based on their propensity to detect a certain class of antibiotics, the LMOFs-based fluorescence sensors discussed. The essay also discusses the field's future trajectory and offers an outlook on the benefits and current difficulties associated with deploying fluorescence sensors based on light metal oxide fragments (LMOFs) for antibiotic detection.

As a colorimetric indicator, the MOF-76 was used to identify several antibiotic species in an aqueous solution (Fig. 1.5). The n-p* transition's intersystem crossing (S1-T1) was made possible by the spin-orbit coupling that the BTC ligand encouraged. By using three distinct detection activities, this luminophore could detect ciprofloxacin (CPFX), nitroimidazole (NDZ), and nitrofurans (NFs) antibiotics numerous times. MOF-76 uses a bathochromic shift of the ligand to identify NFs, luminescence colour changes in CPFX, and lanthanide luminescent quenching to detect NDZ. The MOF76's sensitivity and recyclability were also investigated, with its detection limit remaining at ppb levels for antibiotics [79].

1.4.5 Metal-Organic Frameworks for Detection of Hazardous Chemicals

Environmental contamination is currently one of the most difficult problems on the planet. Numerous experiments have been conducted to lower pollution levels as well as remove harmful substances from the environment. NOx, SOx, COx, H_2S, volatile organic compounds (VOCs), nitrogen-containing compounds (NCCs), sulfur-containing compounds (SCCs), dyes, medicines, and personal care products (PPCPs) are examples of common hazardous components found in our environment. Conversely, the numerous applications of porous metal-organic framework (MOF) materials make them highly intriguing [80, 81]. Because of their high/tunable porosity, pore functionality, diverse pore architectures/compositions, open metal sites, and other features, MOFs are preferable to other porous materials. As a result, a lot of research has been done lately on the adsorption and separation of

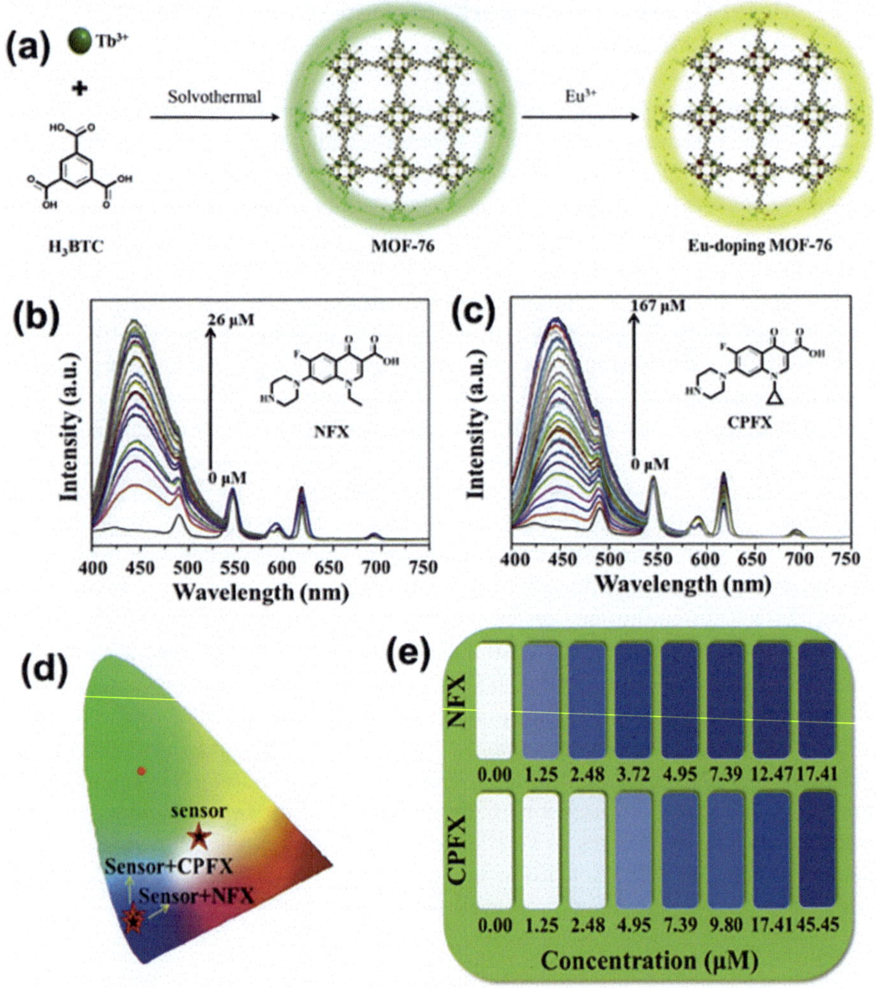

Fig. 1.5 The synthesis of Eu-doped MOF-76 is represented schematically in (**a**); concentration dependence changes in emission spectra for MOF-76 ($Eu_{0.04}Tb_{0.96}$) towards (**b**) NFX and (**c**) CPFX analytes; MOF-76 ($Eu_{0.04}Tb_{0.96}$) before and after mixing with NFX and CPFX is shown in (**d**); and MOF-76 ($Eu_{0.04}Tb_{0.96}$) changes luminescent visually when the concentrations of NFX and CPFX are varied [79]

different liquid and gaseous components using MOFs. This section will cover the use of virgin or modified MOFs for the adsorptive removal of various harmful liquids and gases [82]. An additional instance of reversible, elevated NO_2 uptake was noted in the extremely resilient MFM-300(Al) structure. As Fig. 1.6 illustrates the extraordinary selective elimination of NO_2 at low concentrations (<1 ppm) beyond an absorption capacity of 14.1 mmol/g. Additionally, at ambient temperature and 1.0 pressure, MFM-300(Al) preferentially adsorbed NO_2 with 18.1 over SO_2, 248

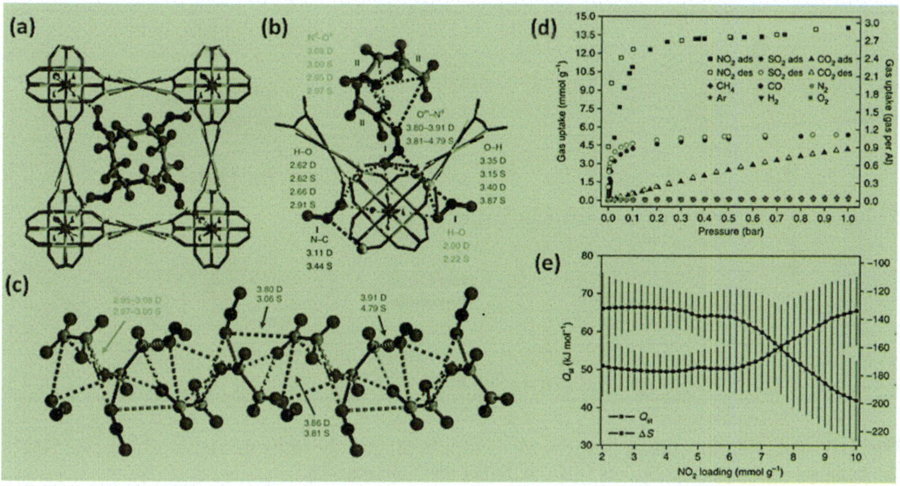

Fig. 1.6 (**a**, **b**) MOF's interactions with the guest molecules in binding sites I and II; (**c**) the formation of a 1D helical chain (NO_2–$2O_4$) in the MFM-300(Al) channel. Bond distances (in Å) were obtained from synchrotron X-ray diffraction (labelled S) and DFT calculations (labelled **d**). Adsorption isotherms of NO2 and several other small gases in MFM-300(Al) at 298 K were also reported, as were (**e**) the isosteric heat of adsorption (Q_{st}) and entropy (ΔS) for NO_2 uptake in MFM-300(Al). Al is green, C is grey, O is red, H is white, and N is blue [83]

over CO_2, and >1000 over N_2. Reversible NO_2 adsorption in the robust framework was validated by density function theory, molecular dynamics calculations, synchronized PXRD, infrared spectroscopy, and EPR spectroscopy. In MFM- 300(Al), a robust interplay between porous hosts and guest molecules was noted through the creation of a $(NO_2{\cdot}N_2O_4)\infty$ monomer–dimer 1D helical chain. Five cycles of NO_2 sorption and desorption also showed MFM-300(Al)'s remarkable structural stability. Following NO_2 adsorption, MFM300(Al) exhibited complete reversibility, enabling the analyte to be collected for usage in a range of possible applications [83].

1.4.6 Metal-Organic Frameworks for Detection of Explosives

The well-defined crystallinity, diverse structural makeup, adjustable porosity, and facilely generated fluorescence emission of luminescent metal-organic frameworks, or LMOFs, have garnered significant interest in the field of MOFs lately. Notably, their features of photoluminescence (PL) can be modified by changing the organic ligands and metal ions in a single hybrid system. One of the most important uses of sensing is the use of LMOFs as chemical sensors to identify explosive and environmentally harmful nitroaromatic chemicals (NACs). The triphenylamine (TPA) unit, which serves as the π-conjugated fluorophores, is the basis for the current developments in the field of NAC detection by LMOFs, which we present in this report [84]. In this regard, Bishwas group [85] reported on the extremely sensitive, selective,

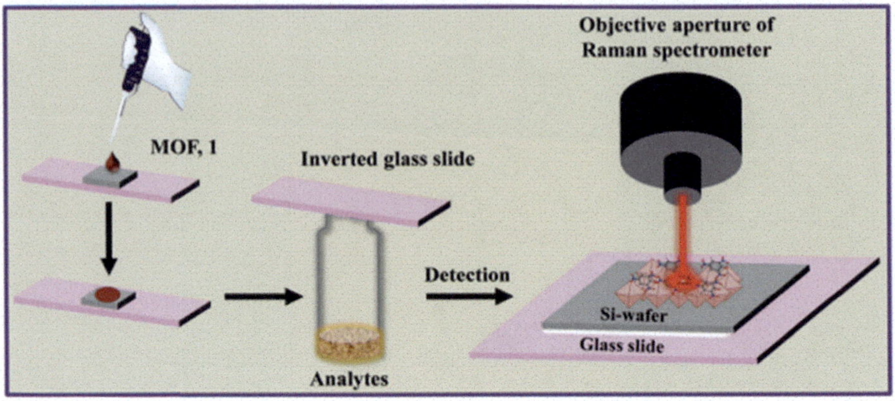

Fig. 1.7 Diagram illustrating how the MOF film is exposed to analyte vapors, which suppresses the peak intensities of the Stokes shift, to detect nitro analytes [85]

fast, and reversible detection of explosive compounds adsorbed on metal–organic frameworks (MOFs) in the vapor phase under usual laboratory circumstances. The method of sensing is based on the fast quenching of 50% of the Raman intensity of a zirconium-based MOF, Zr6O4(OH)4(dcppy)6 (MOF 1, UiO-67-dcppy; dcppy is 2-phenylpyridine-5,40-dicarboxylate), after it is exposed to trinitrophenol (TNP) vapors for B 5 s, and 90% of the raman signal in 30 s. Increased adsorption of different nitro analytes is possible due to the high surface area of porous MOFs. The p–p interactions between the analytes and the MOF's benzene ring caused the dampening of Raman peaks. Additionally, the sensors' reversibility and stability have been demonstrated during a brief, mild heat treatment (80 °C). Given their great adaptability, our noteworthy Raman spectroscopy results suggest a new and significant direction for metal-organic framework explosive sensing. The significance of a non-invasive, non-destructive method for creating portable vapor phase explosive detectors is emphasized by this work (Fig. 1.7).

1.4.7 Metal-Organic Frameworks for Detection of Gunshot Residual

One of the most important pieces of evidence connected to the shooting incidents is gunshot residue, or GSR. GSR analysis makes it easier to investigate important forensic data, such as shooter identification, firefighting range determination, and shooting incident reconstruction. The basis for the GSR study is the propellant charge and primer composition. The traditional GSR examination methodology faced issues once lead-free ammunition was developed through advancements in ammunition manufacture [86]. For the GSR analysis of lead-free ammunition, lanthanide-based metal organic frameworks have proven to be particularly effective as luminous markers. Because of the luminescence it produces, GSR traces can be

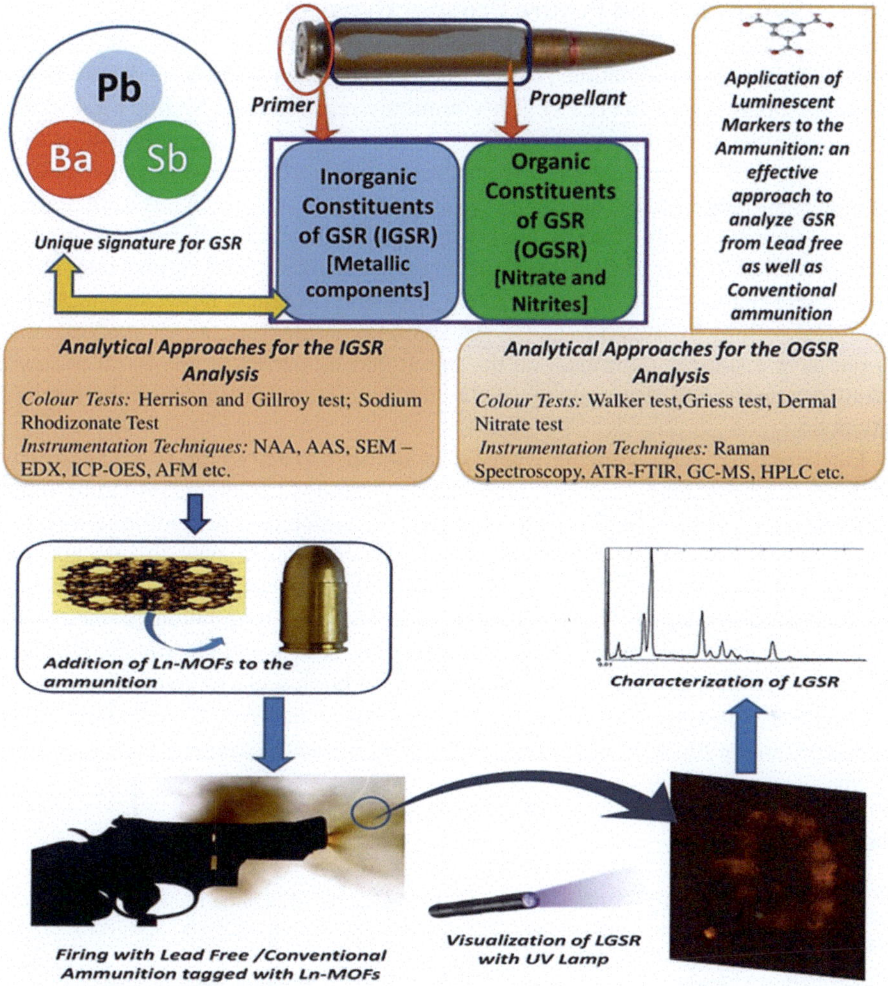

Fig. 1.8 Analytical techniques for GSR analysis [88]

seen with portable UV lights and effectively examined using a variety of methods, including Raman Spectroscopy, SEM-EDX, and the Video Spectral Comparator [87]. Numerous benefits are provided by these fluorescent markers, including easy analysis, improved collecting efficiency, an economical method, etc. (Fig. 1.8). SEM-EDX is regarded as the most precise and widely used technique for GSR analysis among all these others. SEM-EDX detects the distinct particle (Ba, Pb, Sb: components of primer composition) and gives information on the particle's morphology and composition. It appears that the development of GSR markers is opening new avenues for forensic ballistics research. It is imperative to anticipate and promote the use of tagged ammunition in forensic contexts [88].

1.5 Summary and Future Prospectives

A hybrid porous coordination polymer is created when a metal and an organic ligand are bonded together (CPs). It is commonly referred to as MOFs. MOFs have many uses, including the detection of tiny molecules and the creation of luminous indicators. MOFs have a larger response zone than traditional sensors because of their huge surface area and nanoscale porous nature. As a result, it has been discovered that they are highly effective at sensing small compounds. MOFs show promise as markers due to their great thermal and chemical stability. The various interaction phenomena that occur between the framework and the guest molecules in MOFs channels, such as hydrogen bonding, dispersion, coordination, $\pi - \pi$ stacking, and so on, have a substantial impact on the functional properties of MOFs. It places a strong emphasis on choosing the right metal and ligand for the right kind of interaction.

Furthermore, a targeted feature can be accomplished across the bulk material by applying a specific reactive group, chiral center, or redox center and providing functionality to the linkers. Techniques for in situ manufacturing or traditional organic synthesis can have a significant impact on the creation of new linkers and the replacement of commercially available molecules. However, there is a connection between the dimensionality of the final product and the coordination geometry: the size or hardness of the metal center, coordination modes, and electrical configuration all affect the topology of the framework. This book seeks to give a summary of the enhanced MOF-based method's performance in comparison to other, commonly used approaches for various forensic applications as discussed briefly in Sect. 1.4.

References

1. Dwivedi B, Mishra PK, Mishra V. Chapter 19: Chiral metal-organic frameworks for asymmetrical catalysis. In: Metal-Organic Frameworks (MOFs) as catalysts. Springer.) ; ISBN 978-981-16-7958-2; 2022. p. 10–1007.
2. Chatterjee S, Jeevanandham S, Mukherjee M, Vo DVN, Mishra V. Significance of re-engineered zeolites in climate mitigation–a review for carbon capture and separation. J Environ Chem Eng. 2021;9(5):105957.
3. Hu C, Xiao JD, Mao XD, Song LL, Yang XY, Liu SJ. Toughening mechanisms of epoxy resin using aminated metal-organic framework as additive. Mater Lett. 2019;240:113–6.
4. Xing XS, Fu ZH, Zhang NN, Yu XQ, Wang MS, Guo GC. High proton conduction in an excellent water-stable gadolinium metal–organic framework. Chem Commun. 2019;55(9):1241–4.
5. Mirkovic I, Lei L, Ljubic D, Zhu S. Crystal growth of metal–organic framework-5 around cellulose-based fibers having a necklace morphology. ACS omega. 2019;4(1):169–75.
6. Yang H, Bright J, Kasani S, Zheng P, Musho T, Chen B, et al. Metal–organic framework coated titanium dioxide nanorod array p–n heterojunction photoanode for solar water-splitting. Nano Res. 2019;12:643–50.
7. Howlader P, Mukherjee PS. Solvent directed synthesis of molecular cage and metal organic framework of copper (II) paddlewheel cluster. Isr J Chem. 2019;59(3–4):292–8.
8. Yadav N, Mudgal D, Mishra A, Shukla S, Malik T, Mishra V. Harnessing fluorescent carbon quantum dots from natural resource for advancing sweat latent fingerprint recog-

nition with machine learning algorithms for enhanced human identification. Plos One. 2024;19(1):e0296270.

9. Mishra V, Cho JK, Shin SH, Suh YW, Kim HS, Kim YJ. Ruthenium-Na2CO3-catalyzed one-pot synthesis of ring-hydrogenated carbamates from aromatic amines and organic carbonates under H2. Appl Catal A: Gen. 2014;487:82–90.
10. Huan W, Xing M, Cheng C, Li J. Facile fabrication of magnetic metal–organic framework nanofibers for specific capture of phosphorylated peptides. ACS Sustain Chem Eng. 2018;7(2):2245–54.
11. Yang Q, Yang C, Lin C, Jiang H. Metal–organic-framework-derived hollow N-doped porous carbon with ultrahigh concentrations of single Zn atoms for efficient carbon dioxide conversion. Angew Chem. 2019;131(11):3549–53.
12. Rajak R, Saraf M, Mobin SM. Robust heterostructures of a bimetallic sodium–zinc metal–organic framework and reduced graphene oxide for high-performance supercapacitors. J Mater Chem A. 2019;7(4):1725–36.
13. Yang F, Wu M, Wang Y, Ashtiani S, Jiang H. A GO-induced assembly strategy to repair MOF nanosheet-based membrane for efficient H2/CO2 separation. ACS Appl Mater Interfaces. 2018;11(1):990–7.
14. Marshall N, James W, Fulmer J, Crittenden S, Rowe GT. Polythiophene doping of metal-organic frameworks using innate MOF-catalyzed oxidative polymerization. 2018.
15. Yadav N, Mishra V. Organic–inorganic hybrid materials as potential antimicrobial nanocoatings for medical device and implants. In Next-Generation Antimicrobial Nanocoatings for Medical Devices and Implants, Woodhead Publishing, 2024;125–159.
16. Li L, Zhang S, Ruffley JP, Johnson JK. Energy efficient formaldehyde synthesis by direct hydrogenation of carbon monoxide in functionalized metal–organic frameworks. ACS Sustain Chem Eng. 2018;7(2):2508–15.
17. Wang CY, Yu B, Fu H, Wang P, Wang CC. A mixed valence Tb (III)/Tb (IV) metal–organic framework: crystal structure, luminescence property and selective detection of naproxen. Polyhedron. 2019;159:298–307.
18. Wang Q, Liu DJ, Cui LL, Hu XL, Wang XL, Su ZM. A 3D pillared-layer metal–organic framework with fluorescence property for detection of nitroaromatic explosives. New J Chem. 2019;43(2):963–9.
19. Mahmoodi NM, Taghizadeh M, Taghizadeh A. Activated carbon/metal-organic framework composite as a bio-based novel green adsorbent: preparation and mathematical pollutant removal modeling. J Mol Liq. 2019;277:310–22.
20. Mudgal D, Singh RP, Yadav N, Bharti T, Mishra V. Exploring the catalytic efficiency of copper-doped magnetic carbon aerogel for the coupling reaction of isatin oxime with phenylboronic acid derivatives. SynOpen. 2023;04:570–9.
21. Son GM, Truong CC, Mishra DK, Mishra V, Kim YJ. One-pot synthesis of disubstituted urea from carbon dioxide, propylene oxide, and amines catalyzed by imidazolium-tetraiodoindate. Bull Korean Chem Soc. 2018;39(2):174–183.
22. Shen J, Wang N, Wang YG, Yu D, Ouyang X. Efficient adsorption of Pb (II) from aqueous solutions by metal organic framework (Zn-BDC) coated magnetic montmorillonite. Polymers (Basel). 2018;10(12):1383.
23. Esmaeilzadeh M. A composite prepared from a metal-organic framework of type MIL-101 (Fe) and morin-modified magnetite nanoparticles for extraction and speciation of vanadium (IV) and vanadium (V). Microchim Acta. 2019;186:1–7.
24. Zhang L, Li S, Xin J, Ma H, Pang H, Tan L, et al. A non-enzymatic voltammetric xanthine sensor based on the use of platinum nanoparticles loaded with a metal-organic framework of type MIL-101 (Cr). Application to simultaneous detection of dopamine, uric acid, xanthine and hypoxanthine. Microchim Acta. 2019;186:1–10.
25. Mishra V, Kumar R. Cyclic polymer of N-vinylpyrrolidone via ATRP protocol: kinetic study and concentration effect ofpolymer on click chemistry in solution. Polym Sci Ser B. 2019;61:753–61.

26. Yaghi OM, O'Keeffe M, Ockwig NW, Chae HK, Eddaoudi M, Kim J. Reticular synthesis and the design of new materials. Nature. 2003;423(6941):705–14.
27. Dutta A, Pan Y, Liu JQ, Kumar A. Multicomponent isoreticular metal-organic frameworks: principles, current status and challenges. Coord Chem Rev. 2021;445:214074.
28. Hu JS, Lei Z, Pan CL, He J. A new 2D→ 3D polythreaded framework constructed on an N-centered tripodal linker and copper (i). Mendeleev Commun. 2014;24(5):290–2.
29. Hu JS, Zhang XM, Xing HL, He J, Shi JJ. Construction of an unusual 3D framework based on V-shaped imidazolyl and oxalate ligands. Mendeleev Commun. 2013;4(23):231–2.
30. Yutkin MP, Zavakhina MS, Samsonenko DG, Dybtsev DN, Fedin VP. Porous homo-and heterochiral cobalt (II) aspartates with high thermal stability of the metal-organic framework. Russ Chem Bull. 2010;59:733–40.
31. Zavakhina MS, Samsonenko DG, Dybtsev DN, Yutkin MP, Virovets AV, Fedin VP. Synthesis and structure of chiral coordination polymers of Co II, Cu II, and Mg II saccharates. Russ Chem Bull. 2013;62:716–21.
32. Zavakhina MS, Samsonenko DG, Virovets AV, Dybtsev DN, Fedin VP. Homochiral Cu (II) and Ni (II) malates with tunable structural features. J Solid State Chem. 2014;210(1):125–9.
33. Safaei M, Foroughi MM, Ebrahimpoor N, Jahani S, Omidi A, Khatami M. A review on metal-organic frameworks: synthesis and applications. TrAC Trends Anal Chem. 2019;118:401–25. Available from: https://www.sciencedirect.com/science/article/pii/S0165993619301840
34. Chen XY, Zhao B, Shi W, Xia J, Cheng P, Liao DZ, et al. Microporous metal– organic frameworks built on a Ln3 cluster as a six-connecting node. Chem Mater. 2005;17(11):2866–74.
35. Baniani A, Rivera MP, Lively RP, Vasenkov S. Quantifying diffusion of organic liquids in a MOF component of MOF/Polymer mixed-matrix membranes by high field NMR. J Membr Sci. 2021;640:119786.
36. Wu JY, Chao TC, Zhong MS. Influence of counteranions on the structural modulation of silver–di (3-pyridylmethyl) amine coordination polymers. Cryst Growth Des. 2013;13(7):2953–64.
37. Zavyalova AG, Kladko DV, Chernyshov IY, Vinogradov VV. Large MOFs: synthesis strategies and applications where size matters. J Mater Chem A. 2021;9(45):25258–71.
38. Ivanova II, Kolyagin YG. Application of multinuclear MAS NMR for the in situ monitoring of hydrothermal synthesis of zeolites. Chem Eur J. 2021;27(57):14143–67.
39. Chen W, Du L, Wu C. Hydrothermal synthesis of MOFs. In: Metal-organic frameworks for biomedical applications. Elsevier; 2020. p. 141–57.
40. Hu Z, Kundu T, Wang Y, Sun Y, Zeng K, Zhao D. Modulated hydrothermal synthesis of highly stable MOF-808 (Hf) for methane storage. ACS Sustain Chem Eng. 2020;8(46):17042–53.
41. Zhang Y, Bo X, Nsabimana A, Han C, Li M, Guo L. Electrocatalytically active cobalt-based metal–organic framework with incorporated macroporous carbon composite for electrochemical applications. J Mater Chem A. 2015;3(2):732–8.
42. Varsha MV, Nageswaran G. Direct electrochemical synthesis of metal organic frameworks. J Electrochem Soc. 2020;167(15):155527.
43. Sun PP, Zhang YH, Shi H, Shi FN. Controllable one step electrochemical synthesis of PANI encapsulating 3d-4f bimetal MOFs heterostructures as electrode materials for high-performance supercapacitors. Chem Eng J. 2022;427:130836.
44. Campagnol N, Souza ER, De Vos DE, Binnemans K, Fransaer J. Luminescent terbium-containing metal–organic framework films: new approaches for the electrochemical synthesis and application as detectors for explosives. Chem Commun. 2014;50(83):12545–7.
45. MacGillivray LR. Metal-organic frameworks: design and application. Wiley; 2010.
46. Pichon A, Lazuen-Garay A, James SL. Solvent-free synthesis of a microporous metal–organic framework. CrystEngComm. 2006;8(3):211–4.
47. James SL, Adams CJ, Bolm C, Braga D, Collier P, Friščić T, et al. Mechanochemistry: opportunities for new and cleaner synthesis. Chem Soc Rev. 2012;41(1):413–47.
48. Masoomi MY, Morsali A, Junk PC. Rapid mechanochemical synthesis of two new Cd (II)-based metal–organic frameworks with high removal efficiency of Congo red. CrystEngComm. 2015;17(3):686–92.

49. Phang WJ, Lee WR, Yoo K, Ryu DW, Kim B, Hong CS. pH-dependent proton conducting behavior in a metal–organic framework material. Angew Chem. 2014;126(32):8523–7.
50. Jhung SH, Chang JS, Hwang JS, Park SE. Selective formation of SAPO-5 and SAPO-34 molecular sieves with microwave irradiation and hydrothermal heating. Microporous Mesoporous Mater. 2003;64(1–3):33–9.
51. Deeraj BDS, Jayan JS, Raman A, Saritha A, Joseph K. Recent prospects and trends on zeolitic imidazolate frameworks for microwave absorption and EMI shielding applications. Synth Met. 2023;296:117354.
52. Sabouni R, Kazemian H, Rohani S. Microwave synthesis of the CPM-5 metal organic framework. Chem Eng Technol. 2012;35(6):1085–92.
53. Morsali A, Monfared HH, Morsali A, Janiak C. Ultrasonic irradiation assisted syntheses of one-dimensional di (azido)-dipyridylamine Cu (II) coordination polymer nanoparticles. Ultrason Sonochem. 2015;23:208–11.
54. Jung DW, Yang DA, Kim J, Kim J, Ahn WS. Facile synthesis of MOF-177 by a sonochemical method using 1-methyl-2-pyrrolidinone as a solvent. Dalt Trans. 2010;39(11):2883–7.
55. Son WJ, Kim J, Kim J, Ahn WS. Sonochemical synthesis of MOF-5. Chem Commun. 2008;47:6336–8.
56. Haque E, Khan NA, Park JH, Jhung SH. Synthesis of a metal–organic framework material, iron terephthalate, by ultrasound, microwave, and conventional electric heating: a kinetic study. Chem Eur J. 2010;16(3):1046–52.
57. Pineiro-Lopez L, Arcis-Castillo Z, Muñoz MC, Real JA. Clathration of five-membered aromatic rings in the bimetallic spin crossover metal–organic framework [Fe (TPT) 2/3 {MI (CN) 2} 2]· G (MI= Ag, Au). Cryst Growth Des. 2014;14(12):6311–9.
58. Guliani E, Taneja A, Ranjan KR, Mishra V. Luminous insights: Exploring organic fluorescent "Turn-On" chemosensors for metal-ion (Cu^{+2}, Al^{+3}, Zn^{+2}, Fe^{+3}) detection. J Fluoresc 2024; 34(5):1965–2001.
59. Jhung SH, Yoon JW, Hwang JS, Cheetham AK, Chang JS. Facile synthesis of nanoporous nickel phosphates without organic templates under microwave irradiation. Chem Mater. 2005;17(17):4455–60.
60. Truong CC, Mishra DK, Mishra V. Organic carbonate as a green solvent for biocatalysis. In Green sustainable process for chemical and environmental engineering and science, Elsevier, 2021; 253–275.
61. Beamish-Cook J, Shankland K, Murray CA, Vaqueiro P. Insights into the Mechanochemical synthesis of MOF-74. Cryst Growth Des. 2021;21(5):3047–55.
62. Głowniak S, Szczęśniak B, Choma J, Jaroniec M. Mechanochemistry: toward green synthesis of metal–organic frameworks. Mater Today. 2021;46:109–24.
63. Kaupp G. Mechanochemistry: the varied applications of mechanical bond-breaking. CrystEngComm. 2009;11(3):388–403.
64. Behl A, Mishra A, Sharma GP, Mishra V. Detection and determination of the levels of physiologically active substances in non-alcoholic beverages. Int J Med Toxicol. Legal Med. 2021;24(3 and 4):164–74.
65. Willans CE, French S, Anderson KM, Barbour LJ, Gertenbach JA, Lloyd GO, et al. Tripodal imidazole frameworks: reversible vapour sorption both with and without significant structural changes. Dalt Trans. 2011;40(3):573–82.
66. Kulesa KM, Padilha DS, Thapa B, Mazumder S, Losovyj Y, Schlegel HB, et al. A bioinspired cobalt catalyst based on a tripodal imidazole/pyridine platform capable of water reduction and oxidation. J Inorg Biochem. 2023;242:112162.
67. Berger J, Dönmez AS, Ullrich A, Bunzen H, Fischer RA, Kieslich G. Reduced thermal expansion by surface-mounted nanoparticles in a pillared-layered metal-organic framework. Commun Chem. 2022;5(1):177.
68. Rohde MD. Identification of structural phases in pillared-layered, Amino-Functionalized M2 (abdc) 2 (dabco)(M= Co, Ni, Cu, Zn). The University of Bergen; 2023.
69. Bonneau M, Lavenn C, Ginet P, Otake K, ichi, Kitagawa S. Upscale synthesis of a binary pillared layered MOF for hydrocarbon gas storage and separation. Green Chem. 2020;22(3):718–24.

70. Vaitsis C, Sourkouni G, Argirusis C. Sonochemical synthesis of MOFs. In: Metal-organic frameworks for biomedical applications. Elsevier; 2020. p. 223–44.
71. Yu K, Lee YR, Seo JY, Baek KY, Chung YM, Ahn WS. Sonochemical synthesis of Zr-based porphyrinic MOF-525 and MOF-545: enhancement in catalytic and adsorption properties. Microporous Mesoporous Mater. 2021;316:110985.
72. Othong J, Boonmak J, Promarak V, Kielar F, Youngme S. Sonochemical synthesis of carbon dots/lanthanoid MOFs hybrids for white light-emitting diodes with high color rendering. ACS Appl Mater Interfaces. 2019;11(47):44421–9.
73. Khan MM, Rahman A, Matussin SN. Recent Progress of metal-organic frameworks and metal-organic frameworks-based heterostructures as photocatalysts. Nano. 2022;12(16):2820.
74. Zhu Q, Wang W, Sun G, Aryee AA, Wei J, Meng H, et al. Red emissive nanocomposite with high quantum yield for ultrasensitive and selective detection of latent fingerprints. Microchem J. 2023;190:108688.
75. Kumar A, Sahoo SC, Mehta SK, Soni P, Sharma V, Kataria R. A luminescent Zn-MOF for the detection of explosives and development of fingerprints. Anal Methods. 2022;14(7):700–7.
76. Yan L, Yang P, Cai H, Chen L, Wang Y, Li M. ZIF-8-modified Au–Ag/Si nanoporous pillar array for active capture and ultrasensitive SERS-based detection of pentachlorophenol. Anal Methods. 2020;12(32):4064–71.
77. Marimuthu M, Arumugam SS, Sabarinathan D, Li H, Chen Q. Metal organic framework based fluorescence sensor for detection of antibiotics. Trends Food Sci Technol. 2021;116:1002–28.
78. Yin M, Zhang L, Wei X, Sun J, Xu D. Detection of antibiotics by electrochemical sensors based on metal-organic frameworks and their derived materials. Microchem J. 2022;183:107946.
79. Yang Y, Zhao L, Sun M, Wei P, Li G, Li Y. Highly sensitive luminescent detection toward polytypic antibiotics by a water-stable and white-light-emitting MOF-76 derivative. Dyes Pigments. 2020;180:108444.
80. Islamoglu T, Chen Z, Wasson MC, Buru CT, Kirlikovali KO, Afrin U, et al. Metal–organic frameworks against toxic chemicals. Chem Rev. 2020;120(16):8130–60.
81. Woellner M, Hausdorf S, Klein N, Mueller P, Smith MW, Kaskel S. Adsorption and detection of hazardous trace gases by metal–organic frameworks. Adv Mater. 2018;30(37):1704679.
82. Khan NA, Hasan Z, Jhung SH. Adsorptive removal of hazardous materials using metal-organic frameworks (MOFs): a review. J Hazard Mater. 2013;244:444–56.
83. Han X, Godfrey HGW, Briggs L, Davies AJ, Cheng Y, Daemen LL, et al. Reversible adsorption of nitrogen dioxide within a robust porous metal–organic framework. Nat Mater. 2018;17(8):691–6.
84. Shi ZQ, Ji NN, Hu HL. Luminescent triphenylamine-based metal–organic frameworks: recent advances in nitroaromatics detection. Dalt Trans. 2020;49(37):12929–39.
85. Sen BM, Malik M, Poddar P. Raman spectroscopy-based sensitive, fast and reversible vapour phase detection of explosives adsorbed on metal–organic frameworks UiO-67. New J Chem. 2021;45(16):7145–53.
86. Charles S, Geusens N, Nys B. Interpol review of gunshot residue 2019 to 2021. Forensic Sci Int Synergy. 2023;6:100302.
87. de Oliveira Silva JP, Colaço MV, De Souza LA, dos Santos MV, Pugina RS, Marques LF. Exploring the intense lanthanide luminescence and high thermal stability in a new mixed Eu3+/Tb3+ organic framework series for marking in gunshot residues. J Phys Chem C. 2022;126(38):16568–77.
88. Harshey A, Das T, Srivastava A. Analytical contributions of lanthanide based metal-organic frame works as luminescent markers: recent trends in gunshot residue analysis. Microchem J. 2020;154:104597.

Metal Organic Framework: Photochemical and Electrochemical Properties

2

Abstract

This chapter delves into the intriguing realm of Metal-Organic Frameworks (MOFs), focusing on their captivating photochemical and electrochemical properties. MOFs, renowned for their structural diversity and tunable functionalities, offer promising avenues for applications in various domains. The discussion unfolds by elucidating the photochemical behaviors of MOFs, exploring their capacity for light absorption and subsequent photoinduced phenomena such as charge transfer and photoluminescence. By delving into the synergistic interplay between photochemical and electrochemical processes, this chapter unveils the potential of MOFs. Through a concise yet comprehensive examination, this chapter aims to provide insights into leveraging the unique properties of MOFs for advancing diverse applications.

2.1 Introduction

Metal-organic frameworks (MOFs) are an innovative type of material that has completely reshaped the field of the chemical and material sciences. MOFs are porous coordination polymers that are produced by metal-based nodes and multitopic organic ligands or secondary building units (SBUs) through metal–carboxylate or metal–nitrogen coordination bonds [1–3]. MOFs have been used for storage, catalysis, and separation due to their porous and microcrystalline powder structures. Their exceptional structural features are distinguished by metal nodes that are interconnected by organic ligands, assisting in the development of a wide variety of applications, such as drug administration and sensing, gas storage and catalysis. MOFs exhibit a wide variety of functionalities, the most notable of which is their outstanding capacity for detection by photochemical as well as electrochemical processes. Researchers have explored the development of detection platforms that are highly

V. Mishra, A. Mishra, *Metal-Organic Frameworks as Forensic Detectors*,
https://doi.org/10.1007/978-981-96-2147-7_2

sensitive and selective by utilising the intrinsic features of MOFs. This article delves into the intriguing domain of Metal-Organic Frameworks (MOFs) as sensors, with a specific emphasis on their exceptional photochemical and electrochemical characteristics [4–7].

MOFs demonstrate considerable ability as photocatalysts, photoluminescent, light harvester, and light absorbers. The comprehensive engineering of their complicated structure, together with the tunability of its components, facilitates the efficient utilization of light energy. In addition, the porous nature of metal-organic frameworks (MOFs) offers a substantial surface area, hence establishing an optimal setting for catalytic reactions when exposed to light irradiation. However, MOFs exhibit photoluminescent characteristics, rendering them highly beneficial for utilization in sensors and optoelectronic devices. These MOFs may be precisely adjusted to emit light with great accuracy responding to specific requirements [8–10].

MOFs possess exceptional electrochemical stability, rendering them highly suitable for a wide range of applications. The high surface area of these materials enables the effective adaptation of considerable amounts of charge carriers, hence augmenting their potential for energy storage [11, 12]. Furthermore, the presence of catalytic metal sites in select MOFs enables them to serve as efficient electrocatalysts for various reactions, making them pivotal in fuel cells and electrolyzers. In addition, the porous nature of these materials enables effective ion transport, making them appropriate for employing as biosensors, sensors that necessitate elevated ion conductivity, as well as utilization in batteries.

This discussion explores the intersection of MOFs and detection, elucidating their function as sophisticated detectors by focusing on their extraordinary photochemical and electrochemical characteristics. By understanding and implementing, these distinctive properties will enable scientists to create cutting-edge and highly effective detection platforms, promise paradigm-shifting effects on industries including environmental monitoring, healthcare diagnostics, and security applications. A variety of opportunities open as we explore further into the field of MOFs as detectors, providing pathways for ground-breaking improvements in sensing technology.

2.2 MOF as Sensor

Metal-organic frameworks (MOFs) have drawn huge interest in the field of sensors and detectors due to their distinctive structural features and extensive capabilities. These materials, composed of metal ions or clusters coordinated with organic ligands, exhibit exceptional sensitivity and selectivity, making them ideal candidates for detection of substances like gases, pollutants, and biological analytes [10, 13, 14]. The construction of extremely effective and quick-responding sensor systems is made possible by the frequent utilization of detecting methods of MOF, particularly photochemical and electrochemical properties.

2.2.1 Photochemical Properties

MOFs possess fascinating photochemical properties that combine metal ions or clusters with organic ligands. These properties make MOFs promising candidates in the fields of photocatalysis, light harvesting, and photoluminescence [15–17]. Here, we examine into the detailed photochemical properties of MOFs:

2.2.1.1 Photocatalytic Sensors

MOFs can act as efficient photocatalysts due to their large surface area and tunable electronic structure. Certain MOFs, when exposed to light can facilitate a range of chemical reactions, including degradation of organic pollutants, act as efficient photocatalysts and water splitting for hydrogen generation [18–20]. When contaminants or analytes interact with these MOFs under light irradiation, distinctive reactions occur, enabling the sensitive detection of substances like volatile organic compounds (VOCs) and pollutants in air or water. The porous nature of MOFs allows for easy diffusion of reactants to active sites, enhancing their photocatalytic efficiency [18, 21].

Photoluminescence

MOFs exhibit photoluminescence, which means they can emit light when exposed to photons. Photoluminescent metal-organic frameworks (MOFs) can emit distinct wavelengths of light on illumination, and modifications in this emission can serve as an indicator for the existence of analytes. This property arises from the interactions between the metal centers and the organic ligands [22, 23]. By carefully selecting the metal and ligand components, scientists can tune the emission wavelength of MOFs. Functionalization of MOFs with specific ligands allows for selective interaction with target molecules, leading to shifts in luminescence spectra, a phenomenon exploited in fluorescence-based sensors for various applications, including biological and environmental sensing [24].

- *Sensitized Luminescence:*

MOFs have a potential to sensitize luminous materials, enriching their emission properties. For example, this phenomenon is the utilization of MOFs doped with lanthanide ions serves to sensitize the luminescence of these ions, hence resulting in a notable increase in emission efficiency. The utilization of this feature is employed in the design of sensors and imaging agents that exhibit great sensitivity, specifically for applications in the fields of biology and the environment [25].

Light Harvesting

In the context of the detecting process, the term “light harvesting” refers to the procedure of collecting photons (light energy) and transforming them into detectable signals for analytical purpose [26]. MOFs are good materials for light harvesting applications since they can be designed to absorb certain wavelengths of light. By incorporating photosensitive ligands or metal clusters with suitable bandgaps,

MOFs can absorb light energy (photons) and generate excitons (electron-hole pairs). These excitons can be utilized in photovoltaic devices, converting light energy into electrical energy in solar cells [27].

- *Enhanced charge separation and migration:*

Some MOFs can be manufactured such that they have suitable band structures that can facilitate efficient charge separation and migration upon light absorption. The migration process is crucial, as it allows the harvested energy to be concentrated and utilized effectively. Photoexcited electrons and holes can be spatially separated within the MOF structure, preventing recombination, and allowing the harvested energy to be used for various applications, such as in photocatalytic reactions and photodetectors [28, 29].

MOFs in Biosensing for Precise Detection

Among their diverse functionalities, due to the intricate structures and unique properties MOFs stand out, offering a highly sensitive and specific platform for biosensing applications. MOFs have been developed to facilitate accurate detection, in biological and medical applications by harnessing their intrinsic photoluminescent properties [30, 31]. Functionalized MOFs with specific enzymes, ligands or antibodies, target biomolecules (such as proteins, DNA, or pathogens), and interactions occur, inducing changes in the photoluminescent properties of MOF [32]. These changes serve as indicators of the presence or concentration of the target, forming the basis of the biosensing mechanism. MOF photoluminescence can be quenched or enhanced by analytes. When the analyte interacts with the MOF, it quenches its luminescence, while reactions or energy transfers boost it [33]. Here, we explore the fundamental characteristics of MOFs as photoluminescent materials in biosensing:

- *Advantage of MOF in biosensing*

High Sensitivity: MOFs have large surface area and porosity allow biomolecule immobilization, improving sensor sensitivity. MOF-based biosensors can detect trace biomarkers for early illness diagnosis at low biomolecule concentrations [34].

Selective and Specificity: MOF functionalization enhances biosensing selectivity. MOF-based sensors may distinguish closely related analytes by selecting ligands or antibodies that attach to target molecules, assuring detection precision and reliability.

Multiplexing Capability: MOF possess the capability to demonstrate unique photoluminescent reactions in the presence of various analytes. This characteristic facilitates the advancement of multiplexed biosensors having the capability to detect numerous biomolecules. The ability is of great value in intricate biological samples that contain a wide range of analytes [35].

- *Applications in biomedicine*

I. Disease Biomarker Detection: MOF biosensors are utilized in the identification of diseases biomarkers, facilitating the timely detection of a range of ailments such as cancer, infectious diseases, and metabolic disorders. The high specificity and sensitivity of these sensors considerably enhance patient outcomes [36, 37].
II. Pathogen Detection: Biosensors based on MOFs play a pivotal role in the detection and identification of many diseases, including bacteria and viruses. Accurate detection of these pathogens is crucial for the diagnosis of illnesses, maintenance of food safety, and prevention of disease outbreaks.
III. Drug Delivery and Monitoring: MOFs can also be utilized in drug delivery systems. The efficacy of therapies can be enhanced by achieving regulated drug release through the incorporation of therapeutic substances into MOF structures. By incorporating therapeutic agents within MOF structures, controlled drug release can be achieved, enhancing the efficacy of treatments. Additionally, MOF-based sensors can monitor drug levels in the body, allowing for real-time adjustments in dosage, thereby optimizing patient care [38–40].

In conclusion, photoluminescent MOF-based biosensors will revolutionize diagnostics, personalized medicine, and healthcare by detecting biomolecules precisely and quickly to enhance patient outcomes.

Integration of MOF with Other Functional Materials

MOFs can be coupled with other functional materials, such as catalysts and semiconductors, to produce hybrid systems that have improved photochemical characteristics. These hybrid materials take advantage of the benefits offered by both MOFs and the materials into which they are introduced, which results in increased performance in applications involving photocatalysis, sensing, and energy conversion [41–43].

As a conclusion, it can be described that the photochemical properties of MOFs, which include their photocatalytic activity as sensors, photoluminescence, light harvesting capabilities, as biosensor, and integration potential, make them adaptable materials in the field of photonics and photochemistry. These features are still being investigated and improved upon by researchers, which is leading to the creation of new opportunities for novel technologies in the fields of renewable energy, environmental remediation, and sophisticated sensing platforms.

2.2.2 Electrochemical Properties of MOF

MOFs distinguished by their unique porous structure and tunable constituents, have opened the way for novel possibilities in the development of highly efficient electrocatalysts. MOFs based on transition metals (like Co, Ni, Fe, Zn, etc.) are often used as building blocks to make porous carbon, metal-based compounds, and blends. The effective electrochemical processes are facilitated by the porous structure of MOFs, consisting of metal nodes or clusters coupled by organic ligands. MOFs, when utilized as detectors, have numerous benefits within the field of electrochemical sensing [44]. Herein we are examining MOF as electrochemical detectors by harnessing pore structure for efficient sensing:

- *Enhanced surface area and high porosity, structure:*

MOFs feature many active sites for analyte interactions because to their large surface area and porosity. Facilitating the immobilization of electroactive species in a significant quantity ultimately results in an increase in the sensitivity as well as its responsiveness of sensors. High porosity increases ion and molecule movement, making active areas more accessible and promoting efficient electrochemical reactions. MOFs can be synthesized with customized pore sizes and structures to selectively adsorb target molecules for electrode design [45]. Use of some specific MOFs can improve sensor specificity, capturing and detection of desired analytes. Moreover, it is very important to keep these parameters under control for ion transport and to account for the changes in volume that happen when batteries and supercapacitors are charged and discharged [46].

- *Redox active metal centers:*

Some MOFs have metal centers that are redox-active within their frameworks. Because these metal centers can go through reversible redox processes, they are great for finding electrochemical signals that can store and transfer electrons [47, 48]. These redox-active sites make the MOF more electrocatalytic, which makes it easier for electrons to move around during sensing processes. This feature is useful for storing energy, like in batteries and supercapacitors, where MOFs can be used as electrode materials that can hold a lot of charge [49].

- *Ion and molecule transport:*

The porosity of MOFs facilitates the efficient diffusion and transportation of ions and molecules within the material. Ions can move through MOFs that have open metal sites or certain functional groups. These MOFs let ions move in structure, which makes them useful for sensors and batteries in solid-state solutions. The presence of this characteristic plays an essential role in facilitating the fast dispersion of analytes towards the active sites, hence providing speedy response times, and enabling the continuous monitoring of dynamic processes in real-time [50, 51].

- *Stability, reusability, and sensor in severe conditions:*

The stable MOF can be synthesized by utilizing different synthesis techniques. The stable nature of MOF promotes sensor endurance and long-term usage. MOFs are eco-friendly and cost-effective electrochemical detectors because they can be recycled and reused [52, 53]. Since MOF are stable, they can function in harsh environment, especially under high temperatures and pressure conditions. MOF can work in high temperature and pressure conditions since it is stable in nature.

2.2.2.1 Usage of MOF as Electrochemical Detector

(a) **MOF-based electrodes and energy storage application:** MOFs with exceptional electrochemical stability are used as electrode materials. Target analytes react with MOF-coated electrodes, causing modifications in electrical characteristics like current or potential. The detection and quantification of this modification allow for the sensitive and targeted identification of substances like biomolecules, organic chemicals, or heavy metals. MOFs can provide considerable amounts of charge carriers due to their high specific surface area. Although MOFs can be utilized as anode or cathode materials in lithium-ion batteries, this characteristic allowing them to be used as electrode materials in supercapacitors, batteries, and energy storage applications [54–58].
   - *Supercapacitor*

MOFs are enormous for supercapacitor electrodes since they have a high surface area and porosity. They can store charge well at the interface between the electrode and the electrolyte, which makes them a very capacitive charge storage device. MOFs, such as metal-organic frameworks-derived carbon materials (MDCMs), synergistically show exceptional electrochemical performance by combining with the advantages of MOFs and carbon materials [54].

- *Batteries*

MOFs have demonstrated potentially favorable performance in a variety of battery types, particularly lithium-ion batteries (LIBs) and sodium-ion batteries (SIBs). MOFs can function as electrode materials, and they offer high reversible capacities, higher cycle stability, and improved rate capabilities. Metal-organic frameworks (MOFs) have exhibited encouraging outcomes in diverse categories of batteries, including lithium-ion batteries (LIBs) and sodium-ion batteries (SIBs). The capacity to combine wide metal ions into the metal-organic framework (MOF) framework enables the customization of electrodes to suit various battery chemistries [59].

- *Hybrid superconductor and batteries*

MOFs have demonstrated potential for application in hybrid energy storage devices, which offer a synergistic combination of the benefits associated with supercapacitors and batteries. These devices, also referred to as hybrid supercapacitors or lithium-ion capacitors, employ materials based on MOFs as electrodes to attain a harmonious combination of high energy density and high-power density. This characteristic renders them well-suited for a wide range of applications, such as electric vehicles and portable gadgets.

(b) **Electrocatalytic MOFs:** Certain MOFs possess active catalytic sites that enhance electrochemical reactions. These MOFs can serve as efficient electrocatalysts for specific reactions, such as oxygen reduction reaction (ORR) and hydrogen evolution reaction (HER). Integrating these MOFs into sensors enhances their catalytic efficiency, making them valuable for detecting gases or biomolecules based on electrochemical signals [60, 61].

- *Oxygen reduction reaction (ORR) and oxygen evolution reaction (OER).*

MOFs that incorporate metal centers, such as iron, cobalt, or nickel, have been observed to display remarkable catalytic efficacy for the oxygen reduction reaction (ORR) and the oxygen evolution reaction (OER). These reactions play a critical role in the functioning of fuel cells and electrolyzers. The increased surface area and porosity of Metal-Organic Frameworks (MOFs) contribute to a significant increase in active site exposure, hence improving the catalytic performance. The electrocatalytic activity of MOF-derived materials has been enhanced via their development, offering them potentially feasible substitutes for conventional precious metal catalysts [48, 62].

- *Hydrogen evolution reaction (HER)*

MOFs can also serve as efficient catalysts for HER, a crucial process in water electrolysis for the creation of hydrogen. Researchers have successfully enhanced the performance of HER catalysts by integrating precise metal ions or clusters into the framework of MOFs. These catalysts have demonstrated remarkable levels of activity, stability, and scalability. Catalysts produced from metal-organic frameworks (MOFs), commonly in the form of metal sulfides or phosphides, have demonstrated exceptional efficacy in applications involving the hydrogen evolution reaction (HER) [63, 64].

(c) **Ion Conductivity**: MOFs with open metal sites can facilitate ion transport, making them potential candidates for ion-conductive materials in batteries and fuel cells. The porous structure allows for the movement of ions, enhancing the overall performance of these devices [54, 65–67].

- *Sodium-ion batteries (SIBs) and potassium-ion batteries (PIBs)*

MOFs have been extensively investigated as potential electrode materials for sodium-ion batteries (SIBs) and potassium-ion batteries (PIBs). These emerging battery technologies are considered appealing alternatives to lithium-ion batteries (LIBs) primarily because of the abundant availability and relatively low cost of sodium and potassium resources [68]. MOFs have the capability to accommodate sodium or potassium ions within their porous structure, hence serving as reversible sites for ion storage. MOF-based electrodes have shown promising results in terms of cycling stability and high specific capacities, rendering them significant in the advancement of environmentally friendly and economically viable ion batteries [69, 70].

(d) **Sensors and Biosensors**

Metal-organic frameworks (MOFs) exhibit the potential to serve as highly responsive electrodes in electrochemical sensors and biosensors. The considerable surface area of these materials facilitates the effective immobilization of biomolecules or identification elements, resulting in improved sensitivity and selectivity for detecting analytes.

- *Electrochemical sensors*

MOF composites of carbon, nanoparticles, and polymers have the potential to be electrochemical sensors to boost their sensitivity and selectivity. MOF electrodes have been modified with specific ligands, demonstrate remarkable affinity for binding to target analytes. This feature is employed in sensors for the detection of a wide range of molecules, encompassing gases, biomolecules, and heavy metals [71].

- *Enzyme Immobilization*

MOFs offer a reliable and permeable environment for the immobilization of enzymes. MOF may accommodate enzymes inside structures keeping their function active and making them more stable. MOF electrodes that incorporate immobilized enzymes have been widely utilized in the field of biosensors. These electrodes have the advantage of facilitating the detection of biomolecules, such as glucose, cholesterol, and DNA sequences, with a high level of specificity and sensitivity [72, 73].

In summary, MOFs as electrodes offer a versatile platform for energy storage, catalysis, and sensing applications. Their unique properties, combined with the ability to tune their structures and compositions, pave the way for innovative solutions in renewable energy, environmental monitoring, and healthcare. Ongoing research continues to explore the vast potential of MOFs, contributing to the advancement of electrochemical technologies with significant societal and environmental impact.

Research in the field of MOFs is ongoing, and scientists continue to explore new materials and applications. By tailoring the composition and structure of MOFs, researchers can further enhance their photochemical and electrochemical properties, paving the way for innovative technologies in energy conversion, storage, and sensing.

2.3 Hybrid Approaches

(a) **Photoelectrochemical Sensors:** Combining the photochemical and electrochemical properties of MOFs, researchers have developed photoelectrochemical sensors [74]. In these systems, MOFs harvest light energy to create charge carriers, leading to electrochemical reactions with analytes. This synergistic approach results in sensors with improved sensitivity and rapid response times.

(b) **MOF-Integrated Nanomaterials:** MOFs are often integrated with other nanomaterials, such as nanoparticles or nanowires, to enhance detection capabilities. These hybrid structures capitalize on the unique properties of both MOFs and the integrated materials, enabling ultrasensitive and selective detection of target molecules through synergistic interactions [75–77].

MOFs, as detectors, represent a paradigm shift in sensor technology. Their high surface area, tunable structures, and diverse functionalities allow for the creation of sensors with unparalleled performance. By harnessing their photochemical and electrochemical properties, scientists continue to develop innovative sensing platforms that find applications in environmental monitoring, healthcare diagnostics, industrial safety, and beyond. As research advances, the integration of MOFs into sensor technologies is expected to further revolutionize our ability to detect and monitor various substances with exceptional precision and efficiency [78–80].

References

1. Deria P, Mondloch JE, Karagiaridi O, Bury W, Hupp JT, Farha OK. Beyond post-synthesis modification: evolution of metal–organic frameworks via building block replacement. Chem Soc Rev. 2014;43(16):5896–912.
2. Biradha K, Ramanan A, Vittal JJ. Coordination polymers versus metal− organic frameworks. Cryst Growth Des. 2009;9(7):2969–70.
3. Ma X, et al. Recent advances in metal/covalent organic framework-based materials for photoelectrochemical sensing applications. TrAC Trends Anal Chem. 2022;157:116793.
4. Zuliani A, Khiar N, Carrillo-Carrión C. Recent progress of metal–organic frameworks as sensors in (bio) analytical fields: towards real-world applications. Anal Bioanal Chem. 2023;415(11):2005–23.
5. Yadav N, Mudgal D, Mishra A, Shukla S, Malik T, Mishra V. Harnessing fluorescent carbon quantum dots from natural resource for advancing sweat latent fingerprint recognition with machine learning algorithms for enhanced human identification. Plos One. 2024;19(1):e0296270.
6. Rajasree SS, Li X, Deria P. Physical properties of porphyrin-based crystalline metal–organic frameworks. Commun Chem. 2021;4(1):47.
7. Butler KT, Hendon CH, Walsh A. Electronic chemical potentials of porous metal–organic frameworks. J Am Chem Soc. 2014;136(7):2703–6.
8. Wang Z, Huang J, Mao J, Guo Q, Chen Z, Lai Y. Metal–organic frameworks and their derivatives with graphene composites: preparation and applications in electrocatalysis and photocatalysis. J Mater Chem A. 2020;8(6):2934–61.
9. Kitchamsetti N, Chakra CS, De Barros ALF, Kim D. Development of MOF based recyclable photocatalyst for the removal of different organic dye pollutants. Nano. 2023;13(2):336.
10. Yusuf VF, Malek NI, Kailasa SK. Review on metal–organic framework classification, synthetic approaches, and influencing factors: applications in energy, drug delivery, and wastewater treatment. ACS Omega. 2022;7(49):44507–31.
11. Mariyappan S, Agamendran N, Sekar K, Natarajan TS. Metal-organic frameworks (MOFs) for energy production, gaseous fuels and electrochemical energy storage applications. Phys Chem Chem Phys. 2023;(44):25, 30116–30144.
12. Dutt S, Kumar A, Singh S. Synthesis of metal organic frameworks (MOFs) and their derived materials for energy storage applications. Clean Technol. 2023;5(1):140–66.

13. Annamalai J, et al. Synthesis of various dimensional metal organic frameworks (MOFs) and their hybrid composites for emerging applications–a review. Chemosphere. 2022;298:134184.
14. Yi F, Chen D, Wu M, Han L, Jiang H. Chemical sensors based on metal–organic frameworks. ChemPlusChem. 2016;81(8):675–90.
15. di Nunzio MR, Caballero-Mancebo E, Cohen B, Douhal A. Photodynamical behaviour of MOFs and related composites: relevance to emerging photon-based science and applications. J Photochem Photobiol C: Photochem Rev. 2020;44:100355.
16. Ma Q, et al. Recent advances in metal-organic framework (MOF)-based photocatalysts: design strategies and applications in heavy metal control. Molecules. 2023;28(18):6681.
17. Dhakshinamoorthy A, Asiri AM, Garcia H. Metal–organic framework (MOF) compounds: photocatalysts for redox reactions and solar fuel production. Angew Chem Int Ed. 2016;55(18):5414–45.
18. Shi Y, Yang A-F, Cao C-S, Zhao B. Applications of MOFs: recent advances in photocatalytic hydrogen production from water. Coord Chem Rev. 2019;390:50–75.
19. Mishra V, Cho JK, Shin SH, Suh YW, Kim HS, Kim YJ. Ruthenium-Na_2CO_3-catalyzed one-pot synthesis of ring-hydrogenated carbamates from aromatic amines and organic carbonates under H_2. Appl Catal A: Gen 2014;487:82–90.
20. Liu Y, et al. Metal or metal-containing nanoparticle@ MOF nanocomposites as a promising type of photocatalyst. Coord Chem Rev. 2019;388:63–78.
21. Yuan G, et al. MOF-COF composite photocatalysts: design, synthesis, and mechanism. Cryst Growth Des. 2021;22(1):893–908.
22. Yousefi R, et al. MOF-based composites as photoluminescence sensing platforms for pesticides: applications and mechanisms. Environ Res. 2023;226:115664.
23. Deneff JI, et al. Covert MOF-based photoluminescent tags via tunable linker energetics. ACS Appl Mater Interfaces. 2022;14(2):3038–47.
24. Xu X, Ma M, Sun T, Zhao X, Zhang L. Luminescent guests encapsulated in metal–organic frameworks for portable fluorescence sensor and visual detection applications: a review. Biosensors. 2023;13(4):435.
25. Cui Y, Yue Y, Qian G, Chen B. Luminescent functional metal–organic frameworks. Chem Rev. 2012;112(2):1126–62.
26. Lee CY, Farha OK, Hong BJ, Sarjeant AA, Nguyen ST, Hupp JT. Light-harvesting metal–organic frameworks (MOFs): efficient strut-to-strut energy transfer in bodipy and porphyrin-based MOFs. J Am Chem Soc. 2011;133(40):15858–61.
27. Son H-J, et al. Light-harvesting and ultrafast energy migration in porphyrin-based metal–organic frameworks. J Am Chem Soc. 2013;135(2):862–9.
28. Ezugwu CI, Liu S, Li C, Zhuiykov S, Roy S, Verpoort F. Engineering metal-organic frameworks for efficient photocatalytic conversion of CO2 into solar fuels. Coord Chem Rev. 2022;450:214245.
29. Navalón S, Dhakshinamoorthy A, Álvaro M, Ferrer B, García H. Metal–organic frameworks as photocatalysts for solar-driven overall water splitting. Chem Rev. 2022;123(1):445–90.
30. Chandio I, Ai Y, Wu L, Liang Q. Recent progress in MOFs-based nanozymes for biosensing. Nano Res. 2024;17(1):39–64.
31. Yan Y, et al. Metal–organic framework-based biosensor for detecting hydrogen peroxide in plants through color-to-thermal signal conversion. ACS Nano. 2022;16(9):15175–87.
32. Mao H, et al. Recent advances on the metal-organic frameworks-based biosensing methods for cancer biomarkers detection. Crit Rev Anal Chem. 2022:1–17.
33. Ma X, Lepoitevin M, Serre C. Metal–organic frameworks towards bio-medical applications. Mater Chem Front. 2021;5(15):5573–94.
34. Sohrabi H, et al. Metal–organic framework-based biosensing platforms for the sensitive determination of trace elements and heavy metals: a comprehensive review. Ind Eng Chem Res. 2022;62(11):4611–27.
35. Liu Z, Liu Y. Metal– Organic frameworks as sensors of biomolecules. In: Metal– Organic frameworks for environmental sensing. ACS Publications; 2021. p. 1–31.

36. Dong J, Zhao D, Lu Y, Sun W-Y. Photoluminescent metal–organic frameworks and their application for sensing biomolecules. J Mater Chem A. 2019;7(40):22744–67.
37. Al Sharabati M, Sabouni R, Husseini GA. Biomedical applications of metal– organic frameworks for disease diagnosis and drug delivery: a review. Nano. 2022;12(2):277.
38. He S, et al. Metal-organic frameworks for advanced drug delivery. Acta Pharm Sin B. 2021;11(8):2362–95.
39. Maranescu B, Visa A. Applications of metal-organic frameworks as drug delivery systems. Int J Mol Sci. 2022;23(8):4458.
40. Sun Y, et al. Metal–organic framework nanocarriers for drug delivery in biomedical applications. Nano-Micro Lett. 2020;12:1–29.
41. Zhang H, Nai J, Yu L, Lou XWD. Metal-organic-framework-based materials as platforms for renewable energy and environmental applications. Joule. 2017;1(1):77–107.
42. Khan MM, Rahman A, Matussin SN. Recent progress of metal-organic frameworks and metal-organic frameworks-based heterostructures as photocatalysts. Nano. 2022;12(16):2820.
43. Nasalevich MA, Van der Veen M, Kapteijn F, Gascon J. Metal–organic frameworks as heterogeneous photocatalysts: advantages and challenges. CrystEngComm. 2014;16(23):4919–26.
44. Kajal N, Singh V, Gupta R, Gautam S. Metal organic frameworks for electrochemical sensor applications: a review. Environ Res. 2022;204:112320.
45. Farha OK, et al. Metal–organic framework materials with ultrahigh surface areas: is the sky the limit? J Am Chem Soc. 2012;134(36):15016–21.
46. Baumann AE, Burns DA, Liu B, Thoi VS. Metal-organic framework functionalization and design strategies for advanced electrochemical energy storage devices. Commun Chem. 2019;2(1):86.
47. Liu K-K, Meng Z, Fang Y, Jiang H-L. Conductive MOFs for electrocatalysis and electrochemical sensor. eScience. 2023:100133.
48. Peng Y, Sanati S, Morsali A, García H. Metal–organic frameworks as electrocatalysts. Angew Chem. 2023;135(9):e202214707.
49. Monnier V, Odobel F, Diring S. Exploring the impact of successive redox events in thin films of metal–organic frameworks: an absorptiometric approach. J Am Chem Soc. 2023;145(35):19232–42.
50. Wang S, et al. A MOF vertical array enables continuous ion transport pathways with high throughput. J Mater Chem A. 2023;
51. Zeng Q, et al. Cross-linked chains of metal–organic framework afford continuous ion transport in solid batteries. ACS Energy Lett. 2021;6(7):2434–41.
52. Zheng Y, Rao F, Zhang M, Li J, Huang W. Efficient, selective, and reusable metal–organic framework-based adsorbent for the removal of Pb (II) and Cr (VI) heavy-metal pollutants from wastewater. Clean Eng Technol. 2021;5:100344.
53. Xamena FXL, Abad A, Corma A, Garcia H. MOFs as catalysts: activity, reusability and shape-selectivity of a Pd-containing MOF. J Catal. 2007;250(2):294–8.
54. Pettinari C, Tombesi A. MOFs for electrochemical energy conversion and storage. Inorganics. 2023;11(2):65.
55. He B, et al. Freestanding metal–organic frameworks and their derivatives: an emerging platform for electrochemical energy storage and conversion. Chem Rev. 2022;122(11):10087–125.
56. Liu W, Yin X-B. Metal–organic frameworks for electrochemical applications. TrAC Trends Anal Chem. 2016;75:86–96.
57. Jaouen F, Morozan A. Metal-organic frameworks: electrochemical properties. Encycl Inorg Bioinorg Chem. 2011:1–24.
58. Ma T, Li H, Ma J-G, Cheng P. Application of MOF-based materials in electrochemical sensing. Dalt Trans. 2020;49(47):17121–9.
59. Zhao R, Liang Z, Zou R, Xu Q. Metal-organic frameworks for batteries. Joule. 2018;2(11):2235–59.
60. McCarthy BD, Beiler AM, Johnson BA, Liseev T, Castner AT, Ott S. Analysis of electrocatalytic metal-organic frameworks. Coord Chem Rev. 2020;406:213137.

61. Gao J, Huang Q, Wu Y, Lan Y-Q, Chen B. Metal–organic frameworks for photo/electrocatalysis. Adv Energy Sustain Res. 2021;2(8):2100033.
62. Usov PM, et al. Study of electrocatalytic properties of metal–organic framework PCN-223 for the oxygen reduction reaction. ACS Appl Mater Interfaces. 2017;9(39):33539–43.
63. Wang H-F, Chen L, Pang H, Kaskel S, Xu Q. MOF-derived electrocatalysts for oxygen reduction, oxygen evolution and hydrogen evolution reactions. Chem Soc Rev. 2020;49(5):1414–48.
64. Wen X, Guan J. Recent progress on MOF-derived electrocatalysts for hydrogen evolution reaction. Appl Mater Today. 2019;16:146–68.
65. Chai M, et al. Ion transport and conduction in metal-organic framework glasses. J Mater Chem A. 2023;11(38):20302–14.
66. Mehek R, et al. Metal–organic framework based electrode materials for lithium-ion batteries: a review. RSC Adv. 2021;11(47):29247–66.
67. Tran VA, et al. Metal-organic framework for lithium and sodium-ion batteries: progress and perspectives. Fuel. 2022;319:123856.
68. Zhang Y, Sha M, Fu Q, Zhao H, Lei Y. An overview of metal-organic frameworks-derived carbon as anode materials for sodium-and potassium-ion batteries. Mater Today Sustain. 2022;18:100156.
69. Su S, et al. Control of SEI formation for stable potassium-ion battery anodes by Bi-MOF-derived nanocomposites. ACS Appl Mater Interfaces. 2019;11(25):22474–80.
70. Li Y, Zhang J, Chen M. MOF-derived carbon and composites as advanced anode materials for potassium ion batteries: a review. Sustain Mater Technol. 2020;26:e00217.
71. Chen Y, et al. Advanced metal–organic frameworks-based catalysts in electrochemical sensors. Front Chem. 2022;10:881172.
72. Nadar SS, Rathod VK. Magnetic-metal organic framework (magnetic-MOF): a novel platform for enzyme immobilization and nanozyme applications. Int J Biol Macromol. 2018;120:2293–302.
73. Gascón V, Jiménez MB, Blanco RM, Sanchez-Sanchez M. Semi-crystalline Fe-BTC MOF material as an efficient support for enzyme immobilization. Catal Today. 2018;304:119–26.
74. Mo F, Zhou Q, Wang Q, Hou Z, Wang J. The applications of MOFs related materials in photo/electrochemical decontamination: an updated review. Chem Eng J. 2022;450:138326.
75. Dwivedi B, et al., Chiral metal-organic frameworks for asymmetrical catalysis. In Metal-organic frameworks (MOFs) as catalysts, Singapore: Springer Nature Singapore. 2022;493–513.
76. Sheta SM, El-Sheikh SM. Nanomaterials and metal-organic frameworks for biosensing applications of mutations of the emerging viruses. Anal Biochem. 2022;648:114680.
77. Jayaramulu K, et al. Graphene-based metal–organic framework hybrids for applications in catalysis, environmental, and energy technologies. Chem Rev. 2022;122(24):17241–338.
78. Fang X, Zong B, Mao S. Metal–organic framework-based sensors for environmental contaminant sensing. Nano Micro Lett. 2018;10:1–19.
79. Liu L, Zhou Y, Liu S, Xu M. The applications of metal– organic frameworks in electrochemical sensors. ChemElectroChem. 2018;5(1):6–19.
80. Xuan X, Wang M, Manickam S, Boczkaj G, Yoon JY, Sun X. Metal-organic frameworks-based sensors for the detection of toxins in food: a critical mini-review on the applications and mechanisms. Front Bioeng Biotechnol. 2022;10:906374.

Metal-Organic Frameworks (MOFs) for Fingerprint Detection

3

Abstract

This book chapter discusses the novel use of Metal-Organic Frameworks (MOFs) in fingerprint identification, combining materials science, chemistry, and forensics. MOFs' high porosity, adjustable architectures, and various functions make them a suitable platform for fingerprint detecting technology. First, the chapter covers fingerprint analysis basics, highlighting the necessity for sensitive and selective methods in forensic investigations. It then discusses MOFs' customizable nature and large surface area, making them excellent for fingerprint detection. The chapter concentrates on fingerprint-analysis MOF synthesis and design. This involves discussing MOF architectures with functional groups that selectively interact with latent fingerprint components. The chapter also discusses using MOFs as adsorbents to remove molecules from fingerprint residues to improve latent print visibility and persistence. The book chapter also discusses MOF incorporation into fingerprint detecting problems and potential. Scalability, repeatability, and real-world application of MOF-based forensic procedures are considered. This book chapter combines materials science and forensic technology to help researchers, forensic scientists, and professionals improve fingerprint detection techniques. MOFs may improve fingerprint analysis sensitivity and reliability, opening new options for forensic investigations and advancing twenty-first century forensic research.

3.1 Introduction

Fingerprints are distinct and complex patterns that arise from the arrangement of ridges and furrows on the epidermal layer of the skin, mostly found on the tips of the fingers [13]. The examination and use of fingerprints represents a vital component within the field of forensic sciences and biometric identification. The unique

V. Mishra, A. Mishra, *Metal-Organic Frameworks as Forensic Detectors*,
https://doi.org/10.1007/978-981-96-2147-7_3

characteristics shown by these patterns, in conjunction with their enduring presence throughout the course of an individual's lifespan, make fingerprints an indispensable instrument for personal identification and the pursuit of criminal inquiries. The uniqueness of an individual's fingerprints, even those of identical twins, may be attributed to a complex interplay of genetic and environmental variables that shape the establishment of these patterns throughout prenatal development. The identification of fingerprints is based on three main ridge patterns, namely loops, arches, and whorls. These patterns exhibit distinct features known as minutiae, which include ridge endpoints and bifurcations. These minutiae play a crucial role in providing further individualization in fingerprint analysis. Fingerprint analysis is a widely used method employed by forensic professionals to establish connections between prints obtained from crime scenes and those stored in databases, hence facilitating the identification of persons in relation to certain places or actions. Automated Fingerprint Identification Systems (AFIS) are crucial in the effective retrieval and comparison of fingerprints in extensive databases.

The reliability of fingerprint identification within the field of forensic sciences is predicated upon the enduring and distinctive nature of these patterns. Moreover, the use of fingerprint analysis goes beyond the realm of criminal investigations. Fingerprint recognition is extensively used throughout many domains, including access control systems, personal electronic devices such as smartphones, and time and attendance monitoring. Although fingerprint analysis is considered successful, it requires meticulous adherence to set protocols. Additionally, forensic specialists are required to provide testimony in court, elucidating their methodologies and guaranteeing the admissibility of fingerprint evidence. The ongoing incorporation of technical progressions, such as digital databases and enhanced imaging methods, serves to augment the precision and effectiveness of fingerprint recognition, therefore bolstering its continuing importance in the domains of security and personal identity.

The field of research concerned with fingerprints is often referred to as "Dactyloscopy," which demonstrates the distinctiveness of fingerprints as an individual's unique trait. Each individual carries a distinct fingerprint. The pattern of a fingerprint is consistent throughout the course of an individual's lifetime; however, changes in the characteristics of the print might occur due to chronic illnesses and permanent scarring. Fingerprints are differentiated based on the ridge features that are visible on their surface. Every fingerprint is comprised of a pattern of ridges and troughs. The ability to determine the discriminating capacity of a fingerprint is achieved by examining the minutiae points, which represent specific sites where the ridge structure undergoes changes, together with the overall pattern of ridges and valleys (Fig. 3.1).

The term "points of similarity" or "points of identification" is used to describe certain characteristics that align between two separate prints. At now, there is a lack of consensus over the minimal threshold of identification points required to create a conclusive match between two fingerprints. In contrast, Australia enforces a

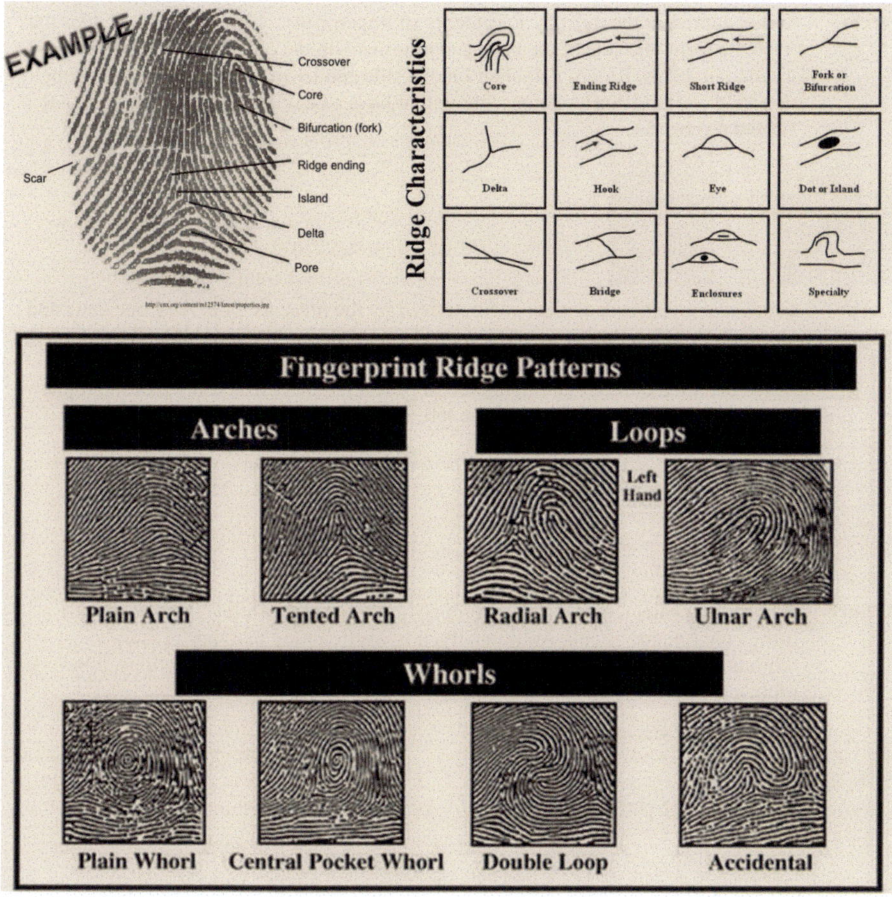

Fig. 3.1 Different Fingerprint Ridge patterns, and characteristics. http://6enders.weebly.com/fingerprints.html; https://www.crime-scene-investigator.net/FingerprintRidgePatternsAndCharacteristics.html

minimum requirement of 12 points, while the United Kingdom maintains a stricter standard of 16 points.

3.1.1 Fingerprint Patterns and Classes

Fingerprints may be categorized into three unique types, namely arches, loops, or whorls, based on their visual pattern. Consequently, it is possible to further break each of these groups into smaller clusters.

Arches	Arches represent the most basic category of fingerprints, characterized by ridges that originate from one side of the print and terminate on the other side. There is an absence of deltas. Ridges that enter on one side and leave on the other side may be classified as **Plain Arch**, whereas those that have a central spike are categorized as **Tented Arch**.
Plain Arch 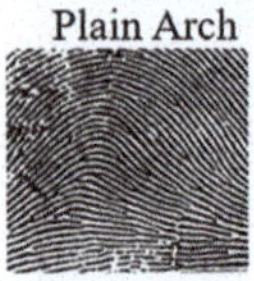	Tented Arch 
Loops	Loops in a given structure are characterized by the presence of at least one delta and one or more ridges that both originate and terminate on the same side. The nomenclature of these patterns is derived from their spatial relationship with the radius and ulna bones. The orientation of the loop may either be towards the right, namely the ulna bone, or towards the left, specifically the radial bone.
Ulnar Loop (Right Thumb) 	Radial Loop (Right Thumb) 
Whorls	Whorls are characterized by the presence of a ridge that exhibits the tendency to form a continuous circuit. Additionally, it is worth noting that they possess a minimum of two deltas. If a print exhibits more than two delta formations, it is very probable that it is an incidental occurrence.
Plain Whorl 	Central Pocket Whorl 

3.1.2 Metal-Organic Frameworks for Fingerprint Detection

Metal-Organic Frameworks (MOFs) have become a novel class of materials with a wide range of scientific uses. These extremely adaptable and porous materials are made up of organic ligands in coordination with metal ions or clusters, creating three-dimensional structures with remarkable surface areas and chemical characteristics that may be adjusted [14]. MOFs have received a lot of interest lately in the forensic science community, particularly with regard to fingerprint analysis and detection [1]. The porous nature of MOFs provides an ideal environment for the adsorption and capture of various compounds, including amino acids, peptides, and other organic molecules present in fingerprints. MOFs can be engineered to have specific functionalities and tailored pore sizes, enabling them to selectively bind to target compounds found in latent fingerprints. This selectivity is vital for enhancing the sensitivity and accuracy of fingerprint detection methods. MOFs exhibit a porous structure, making them highly suitable for a wide array of applications including gas storage, catalysis, and sensing. The technology known as surface-enhanced Raman scattering (SERS) has undergone significant advancements and emerged as a very effective spectroscopic method for fingerprint analysis [2]. This technique offers several notable advantages, including its ability to detect chemical substances swiftly and accurately without causing harm or damage. Furthermore, SERS exhibits an exceptional level of sensitivity, capable of detecting even individual molecules [1]. As a result, many researchers paired MOFs with SERS to enhance sensing capabilities and increase the range of applications for SERS substrates [1]. MOFs have been employed as substrates for ultra-sensitive biological trace detection using surface-enhanced Raman spectroscopy (SERS) [1, 3]. A summary of typical MOF synthesis methods for creating SERS-active substrates is provided, along with a discussion of its uses in ultra-sensitive biological trace detection. A detailed discussion is held regarding the fundamentals of detecting biomolecules using lanthanide complexes and lanthanide up conversion nanoparticles (UCNPs) [4]. MOFs provide an immense amount of potential for fingerprint identification, especially when coupled with SERS technology. Their synthesis methods and applications have been thoroughly investigated, and they have been employed as SERS-active substrates for ultra-sensitive biological trace detection.

A key component of forensic investigations is fingerprint detection, which helps law enforcement identify people and solve crimes. There are drawbacks to conventional fingerprint detection techniques, such as powder dusting and chemical reagent use, including issues with sensitivity, selectivity, and environmental impact. However, distinct qualities of MOFs make them a possible alternative.

In this sequence of detection, recently in 2023, Wiwasuku et al. was synthesized a water-stable Europium Metal-Organic Framework (Eu-MOF) through a successful one-step hydrothermal process, featuring free carboxyl functional groups for effective interaction with guest molecules. This Eu-MOF served as a dual-responsive fluorescent sensor, demonstrating its versatility as a turn-on probe for tetracycline antibiotics and a turn-off probe for dihydrogen phosphate. Notably, the lower limits of detection for OTC, TC, and DOX were determined as 78 nM, 225 nM, and 201 nM, respectively. The observed enhancement of Eu-MOF by tetracycline antibiotics was attributed to hydrogen bond and coordination interactions facilitating energy and electron transfer from antibiotics to the MOF. Additionally, the Eu-MOF exhibited high sensitivity and selectivity in detecting $H_2PO_4^-$ ions in aqueous solutions, with a low limit of detection of 0.70 mM. The fluorescent quenching phenomenon [15] resulted from the interference caused by the interaction between Eu-MOF and $H_2PO_4^-$ during the antenna process. Crucially, the sensor demonstrated applicability in authentic samples, including milk, chicken breast, honey, and water, with good recovery rates. These results underscore Eu-MOF's potential as a sensing material for efficient identification of antibiotics and anions. This research contributes to the design of a highly effective dual-responsive fluorescence sensor based on MOFs, promising applications in food safety and environmental protection. Moreover, the Eu-MOF proved valuable in developing latent fingerprints, producing dependable images with high contrast, minimal interference, and user-friendly application. The anticipated Eu-MOF holds significant potential for diverse

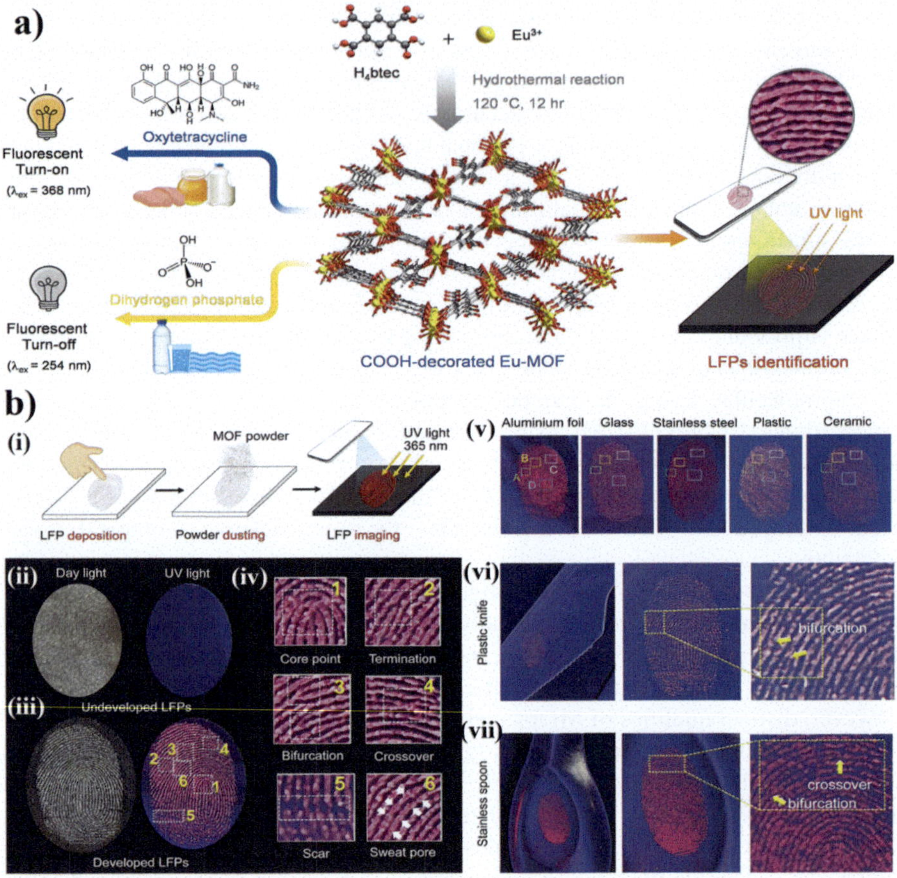

Fig. 3.2 MOF for fingerprint detection. (**a**) Scheme diagram for synthesis of water-stable Eu-MOF and its application for the detection of oxytetracycline and $H_2PO_4^-$ and visualization of latent fingerprints (LFPs) (**b**) (i) Steps for using MOF powders for fluorescent LFPs visualization. Photographs of undeveloped (ii) and developed (iii) latent fingerprints on glass slide under day light and UV light (365 nm). (iv) Enlarged images showing level 1–3 features of fingerprint details as marked in (**c**) including (1) core point, (2) termination, (3) bifurcation, (4) crossover, (5) scar, and (6) sweat pore. (v) Photographs of latent fingerprints developed with EuMOF materials on different substrates under UV light. The marked areas show (A) termination, (B) bifurcation, (C) crossover, and (D) core point. (vi–vii) Photographs of latent fingerprints developed with Eu-MOF on the surface of a plastic knife (vi) and a stainless spoon (vii) under UV light. The left and middle sets of images show visualized latent fingerprints under 365 nm UV light. The right set of images show bifurcation and crossover on the latent fingerprints. (Reprint with permission from Reference [5]. Copyright@2023, RSC publications)

applications in the display of latent fingerprints (LFPs). The diagram in Fig. 2a demonstrates the standard procedure for synthesizing a water-stable fluorescent sensor using Eu-MOF. This sensor is capable of detecting and visualizing oxytetracycline and $H_2PO_4^-$ in solution, as well as facilitating the visualization of LFPs [Fig. 3.2] [5]. The method for fluorescent imaging of local field potentials is shown in Fig. 3.1b (i). The observation of the intricacies and patterns of developing fingerprints on the glass slide was hindered by insufficient clarity under both daylight and 365 nm UV light, as seen in Fig. 3.2b (ii). Ideally, the technology being developed for latent fingerprint processing should include all the unique features associated with fingerprints, ranging from level 1 characteristics such as the general shape of the fingerprint and the core point, to level 2 features like bifurcation, termination, and crossing, and finally to level 3 attributes like scars and sweat pores. Upon the application of the Eu-MOF powder onto the substrate, the pattern feature at level 1 of the produced fingerprint was promptly and clearly noticed when exposed to UV light. This observation revealed a significant enhancement in contrast between the fluorescent ridges and the non-fluorescent furrow, as shown in Fig. 3.2b (iii). Furthermore, in addition to this, the magnified fingerprint also allows for the observation of information at both the core point (level 1) and termination, bifurcation, and crossing points (level 2) without the need of any magnifying devices. Interestingly, it is possible to discern level 3 features, such as scars and sweat pores, without the need for any posttreatment methods, as seen in Fig. 3.2b (iv). The level 2 and 3 features provide an adequate quantity of individual information, making them suitable for fingerprint analysis.

The branch of fingerprint detection necessitates ongoing advancements in detection techniques to enhance detection thresholds, enabling the identification of a greater number of fingermarks with enhanced precision. The techniques should possess the attributes of being efficient in their application, cost-effective, and easily accessible to users. Furthermore, it is imperative that these procedures, when implemented consecutively, demonstrate compatibility with other frequently employed methods and facilitate the utilization of other trace sources, such as DNA [6].

Furthermore, MOFs can be integrated into advanced sensing platforms and imaging techniques, allowing for the development of highly sensitive and rapid fingerprint detection systems. These systems not only improve the efficiency of forensic investigations but also contribute to the advancement of forensic science.

In this context, this introduction explores the innovative applications of Metal-Organic Frameworks in the realm of fingerprint detection, highlighting their potential to revolutionize the field by providing more reliable, sensitive, and selective techniques for identifying individuals and solving crimes. Through ongoing research and development, MOFs continue to pave the way for the future of forensic analysis, ensuring the continued evolution of techniques used in law enforcement and criminal justice. Here's how MOFs work for fingerprint detection:

3.2 How MOF Detects Fingerprints?

1. *Adsorption of Organic Compounds:* MOFs exhibit a significantly porous architecture, characterized by a substantial surface area and distinct pore dimensions. The aforementioned pores have the capacity to adsorb chemical substances that are commonly found in fingerprints, including amino acids, peptides, and fatty acids. The customized structure of MOFs enables them to exhibit preferential adsorption properties towards specific organic molecules [7].

2. *Chemical Affinity:* MOFs possess the capability to be deliberately modified in order to exhibit chemical capabilities that preferentially interact with the organic constituents present in fingerprints. MOFs have the potential to demonstrate a significant affinity for the specific chemicals present in fingerprints through the careful selection of appropriate metal ions and ligands. The chemical affinity exhibited by MOFs plays a crucial role in their ability to selectively capture specific components, while effectively rejecting other chemicals that may be present on surfaces [8].
3. *Selective Binding:* MOFs exhibit the ability to preferentially adhere to specific molecules by virtue of their size, shape, and chemical characteristics. The process of selective binding plays a critical role in distinguishing between various compounds within intricate samples, such as latent fingerprints found on diverse surfaces. MOFs possess the capability to be intentionally engineered in order to effectively sequester particular molecules, hence augmenting the selectivity of fingerprint detection techniques [9].
4. *Enhanced Sensitivity:* The utilization of MOFs in fingerprint detecting techniques is attributed to their notable characteristics, including their substantial surface area and unique interactions. These attributes contribute to the heightened sensitivity observed in the detection process. Trace amounts of organic substances can be captured by these devices, so enabling the detection of fingerprints even under difficult settings when conventional techniques may prove ineffective [1, 10].
5. *Integration into Detection Systems:* MOFs provide the capability to be seamlessly included into a wide range of sensing systems, encompassing sensors and imaging methodologies. MOFs, when utilized in conjunction with suitable transducers or analytical instruments, enable the advancement of highly sensitive and expeditious fingerprint detecting systems. These systems have the capability to do real-time analysis and generate precise outcomes, thereby assisting forensic investigators in the identification of individuals through the examination of their fingerprints [11, 12].
6. *Preservation and Archiving:* MOFs also possess the notable benefit of effectively retaining fingerprints [13] for extended periods of time. The organic components that have been captured can be effectively preserved within the structure of the MOF, so ensuring the long-term preservation of the integrity of the fingerprint samples. The preservation component of this phenomenon holds significant value in the context of forensic investigations, as it enables the possibility of reanalysing preserved samples, if the need arises [13].

References

1. Zhang Y, Xue C, Xu Y, Cui S, Ganeev AA, Kistenev YV, Gubal A, Chuchina V, Jin H, Cui D. Metal-organic frameworks based surface-enhanced Raman spectroscopy technique for ultra-sensitive biomedical trace detection. Nano Res. 2023;16:2968–79.
2. Gao S, Guo Z, Liu Z. Recent advances in rational design and engineering of signal-amplifying substrates for surface-enhanced Raman scattering-based bioassays. Chemosensors. 2023;11:461.

3. Fang X, Zong B, Mao S. Metal–organic framework-based sensors for environmental contaminant sensing. Nano Micro Lett. 2018;10:1–19.
4. Sun G, Xie Y, Sun L, Zhang H. Lanthanide upconversion and downshifting luminescence for biomolecules detection. Nanoscale Horiz. 2021;6:766–80.
5. Wiwasuku T, Chuaephon A, Puangmali T, Boonmak J, Ittisanronnachai S, Promarak V, Youngme S. Multifunctional fluorescent Eu-MOF probe for tetracycline antibiotics and dihydrogen phosphate sensing and visualizing latent fingerprints. RSC Adv. 2023;13:10384–96.
6. Moret S, Scott E, Barone A, Liang K, Lennard C, Roux C, Spindler X. Metal-organic frameworks for fingermark detection—a feasibility study. Forensic Sci Int. 2018;291:83–93.
7. Chen L, Xu Q. Metal-organic framework composites for catalysis. Matter. 2019;1:57–89.
8. Raptopoulou CP. Metal-organic frameworks: synthetic methods and potential applications. Materials (Basel). 2021;14:310.
9. He S, Wu L, Li X, Sun H, Xiong T, Liu J, Huang C, Xu H, Sun H, Chen W. Metal-organic frameworks for advanced drug delivery. Acta Pharm Sin B. 2021;11:2362–95.
10. Xiong W, Wang X, Liu H, Zhang Y. Metal–organic frameworks–based surface–enhanced Raman scattering substrates for gas sensing. Chemosensors. 2023;11:541.
11. Shen Y, Tissot A, Serre C. Recent progress on MOF-based optical sensors for VOC sensing. Chem Sci. 2022;13:13978–4007.
12. Mukherjee S, Ghosh S, Biswas S. A MOF chemosensor for highly sensitive and ultrafast detection of folic acid in biofriendly medium, paper strips and real samples. Inorg Chem Front. 2022;9:6288–98.
13. Yadav N, Mudgal D, Mishra A, Shukla S, Malik T, Mishra V. Harnessing fluorescent carbon quantum dots from natural resource for advancing sweat latent fingerprint recognition with machine learning algorithms for enhanced human identification. Plos One. 2024;19(1):e0296270.
14. Dwivedi B, Mishra PK, Mishra V. Chiral metal-organic frameworks for asymmetrical catalysis. In metal-organic frameworks (MOFs) as catalysts, Singapore: Springer Nature Singapore, 2022;493–513.
15. Guliani E, Taneja A, Ranjan KR, Mishra V. Luminous insights: Exploring organic fluorescent "Turn-On" chemosensors for metal-ion (Cu+ 2, Al+ 3, Zn+ 2, Fe+ 3) detection. J Fluoresc. 2024;34(5):1965–2001.

MOF for Lip Print and Palm Print Detection

4

Abstract

Metal-organic frameworks (MOFs) have received considerable attention in the scientific community owing to their intriguing structural properties and wide variety of uses. MOFs provide the capability to boost selectivity towards specific analytes by responding to various chemical and physical stimuli. In this chapter, a detailed study was presented about the usage of MOF for the detection of lip and palm print. The utilization of MOF based colorimetric or fluorescent sensors presents a visually perceptible and readily interpretable approach for the detection of lip and palm prints. Interestingly, MOF can create sensitive detecting methods to selectively interact with organic molecules that are present in human lip print and can also be constructed with high affinity to absorb VOCs to concentrate and isolate trace components in human palm prints. A framework of rare earth metals with lanthanides and transition metals are designed that are fluorescent in nature and can work effectively for lip print and palm print detection.

4.1 Introduction

An individual's identity represents a unique and exclusive set of characteristics and attributes that are specific to each person. It anticipates an essential role in instances of theft, criminal inquiries, and mass tragedies. Previously, fingerprints [1], DNA, and retina patterns have been utilized for the specific goal of identifying individuals [2, 3].The discipline of forensic science has witnessed several advances due to the rapid progress in technology. While there are several established techniques for human identification [1]. such as DNA comparisons and dental fingerprints, one particularly noteworthy developing method is human lip recognition. The lip's outer surface has many protuberances and indentations, together known as lip prints, which possess a distinct configuration exclusive to each person [3, 4]. Lip print

V. Mishra, A. Mishra, *Metal-Organic Frameworks as Forensic Detectors*,
https://doi.org/10.1007/978-981-96-2147-7_4

identification has been proposed as an additional technique for criminal investigation, based on the claimed uniqueness of lip grooves. Nevertheless, critics have raised doubts over the reliability and uniformity of the methodologies and procedures employed, necessitating the development of updated standards and criteria to validate this field as a bona fide forensic identification science. Cheiloscopy is a forensic method used to identify individuals by analyzing lip traces. Lip prints, similar to fingerprints, possess distinct and enduring qualities that are exclusive to each person [5, 6]. Additionally, the creases on the lips exhibit unique features for each individual. Cheiloscopy is regarded as a tool for individual identification and might serve as corroborating evidence in particular investigations [7]. It has resemblance to fingerprint analysis and is a legitimate branch of forensic odontology. Lip prints possess distinctive characteristics that facilitate the identification of individuals [8]. This approach is valuable in forensic sex determination and criminal investigation and serves as a tool for personal identification [9]. The outer layer of the lip has several raised and lowered areas that create a distinct pattern known as lip prints. The analysis of these prints is known as cheiloscopy. This is distinct to people, much like fingerprints. Anthropologist R.S. Fischer first observed and documented the biological occurrence of furrow systems on the red portion of human lips in lips 1902, acknowledged by Kasprzak in 1990 [10]. However, prior to 1930, anthropology just acknowledged the presence of furrows without proposing a practical use for this phenomenon. Since 1950, the Japanese have conducted thorough study on this subject. Between 1968 and 1971, Y.T Suchhihashi and T. Suzuki conducted a study at the department of forensic odontology at Tokyo University [11]. They investigated 1364 individuals and determined that the pattern of lines on the red region of the human lips is distinct and exclusive to each person. Lip print recording is a valuable tool in forensic inquiry for identifying individuals through the analysis of lip prints. A lip print can be observed as a surface displaying visible lines that correspond to the furrows [12]. This distinctive pattern aids in the identification of individuals as it is exclusive to each individual [13].

Suzuki and Tsuchihashi (1968–1971) conducted research that confirmed the distinct and unique arrangement of lines on the red region of human lips for each person [14, 15]. This phrase implies that the pattern of furrows on lips may potentially be used to identify an individual [16].

The uniformity and permanence of Lip Prints make them a reliable means of identifying individuals at crime scenes, since these distinct patterns serve as distinctive markers [17]. Lip prints taken from lips that are not covered with lipstick are classified as latent prints. Latent lip prints are regarded as a crucial instrument in criminal identification [18]. The latent lip prints were effectively created for investigation purposes utilizing different colors [19].

Here are the steps involved in conducting cheiloscopy:

1. **Lip print development**: The subject (participant) is instructed to press their lips (without using lipstick or any other substance for marking) on an appropriate surface, such as paper or glass [20].

2. **Powder application**: To enhance the visibility of the lip print, an appropriate powder, such as black carbon powder or fluorescent powder is utilized [21, 22].
3. **Lip print lifting**: With the purpose to lift-off the lip print from the surface, a piece of clear adhesive tape, such as cellophane tape, is applied to the lip print [23, 24].
4. **Magnification**: The elevated lip print is scrutinized using a magnifying lens to identify the distinct features of the wrinkle pattern on the lips [25].
5. **Comparison**: A comparison is made between the lifted lip print and the known lip prints of the individual to determine a match [26].
6. **Reporting**: The findings are documented and assessed to ascertain the unique characteristics of the lip print and its significance as testimony in the investigation.

Lip prints found at crime scenes are often overlooked by detectives and crime scene examiners. It is worth emphasizing that lip prints found at crime scenes are more common than people often believe. Lip prints may be detected on a range of surfaces including glass, plates, clothing, utensils, and cigarette butts [17, 27–29]. Likewise, the imperceptible or dormant lip prints can also be shown by employing aluminum and magnetic powder. The effectiveness of developing lip prints using various chemical developers relies on the surface of the item where the lip print was formed and the nature of the chemical compound employed [19, 30, 31].

4.1.1 Powder Sources for Lip-Print Identification

Several types of powders may be utilized for lip print identification, such as fingerprint black powder, aluminum powder, silver metallic powder, silk black "Hi-Fi" volcano latent print powder, and white volcanic latent print powder. The powders are put to the surface where the lip print is believed to exist, and any extra powder is eliminated to reveal the concealed print [32, 33]. The print can subsequently be transferred onto a white bond sheet using lifting tape. Additional powders that have been utilized include Sudan III, indigo, and traditional white powder. The efficacy of the powder utilized relies on the nature of the surface on which the lip print is discovered, as well as the caliber of the print itself. Adhering to a rigorous process is crucial while monitoring and obtaining photographic photographs of the prints, regardless of the absence of any lipstick residue [1, 34, 35] (Fig. 4.1).

It was found that lysochrome dyes, including Sudan III, aluminum powder, indigo dye, and fingerprint powder, have been shown effective for developing latent lip prints [31, 32, 36]. The study determined that fingerprint powder was the most effective developer, followed by Sudan III and vermilion. In previous studies, fingerprint powder was recognized as highly significant due to its composition and mechanical characteristics. In a study conducted by Alvarez Segui et al. in 2000, it was shown that black and silver metallic powders are the most efficient at revealing hidden lip prints [18, 37]. However, Trozzi et al. (2001) concluded that Sudan black is the most effective ingredient for enhancing lip prints that have been polluted with foodstuff, oils, and other fatty substances [38]. In contrast, Alvarez and Castello

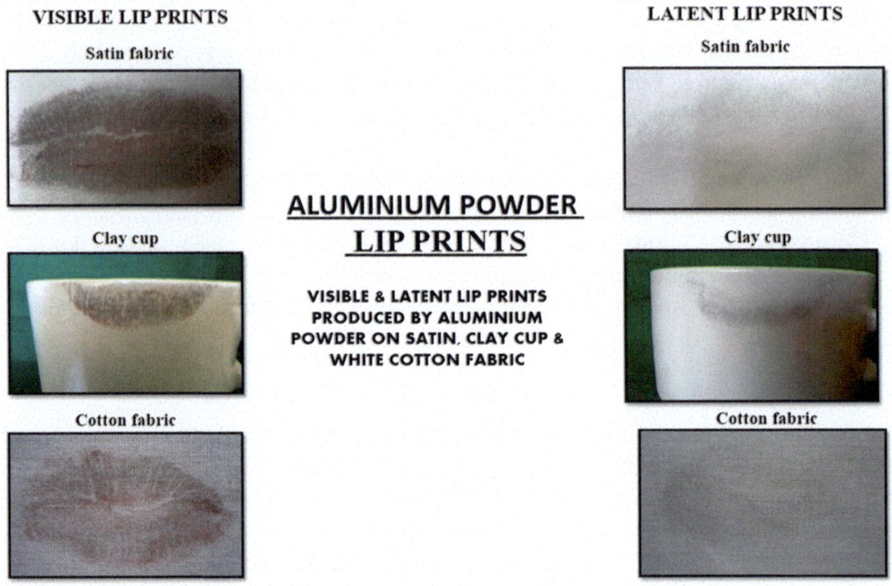

Fig. 4.1 Different materials used in cheiloscopy [32]

(2002) discovered that lysochrome dyes are superior than fingerprint powder for creating lip prints [39]. Black fingerprint powder is versatile in its use due to its small particle size and strong mechanical affinity to latent print residue. The Sudan III powder is exclusively suitable for achieving smooth and shiny surfaces. The vermilion powder exhibited a high density and irregular distribution, resulting in an inability to capture the fine details of the latent lip print [37–40]. MOFs are a category of materials distinguished by their distinctive architecture, comprising metal ions or clusters that are coordinated with organic ligands [85]. MOFs possess a very porous structure, rendering them exceptionally well-suited for a diverse range of applications such as gas storage, catalysis, and sensing [41]. In recent years, scholars have conducted investigations into the application of MOFs within the field of forensics, specifically focusing on the identification and examination of lip prints. In the field of materials science looked at how MOFs can be grown on proteins that have been printed on a surface. The classic powder dust approach for fingerprints has poor sensitivity, low resolution, and a lot of background interference. Smoking, vacuum coating, and other procedures for lip lines have poor visibility and a complicated operation. The field of Rare-earth-doped luminous materials, such as LaOF: Sm^{3+}, $CaAl_2O_4$: Eu^{2+}, Dy^{3+}, MoO_3:Dy^{3+}, and $CaAl_2O_4$:Tb^{3+}, Ce^{3+}, M^+ (M^+ = K, Na, Li), have recently been extensively employed in luminescence and have a startling impact in identifying putative fingerprints and lip lines at crime scenes [42–44]. Specifically, lanthanide doped are being utilized for possible lip image; lanthanide doped rare earth nano ions have strong optical contrast and high detection sensitivity and are scarcely influenced by background fluorescence (Fig. 4.2).

Lip prints are unique to individuals and are uniform throughout life; it can be classified into five types (Fig. 4.3). Lip prints are convenient in criminal science, the

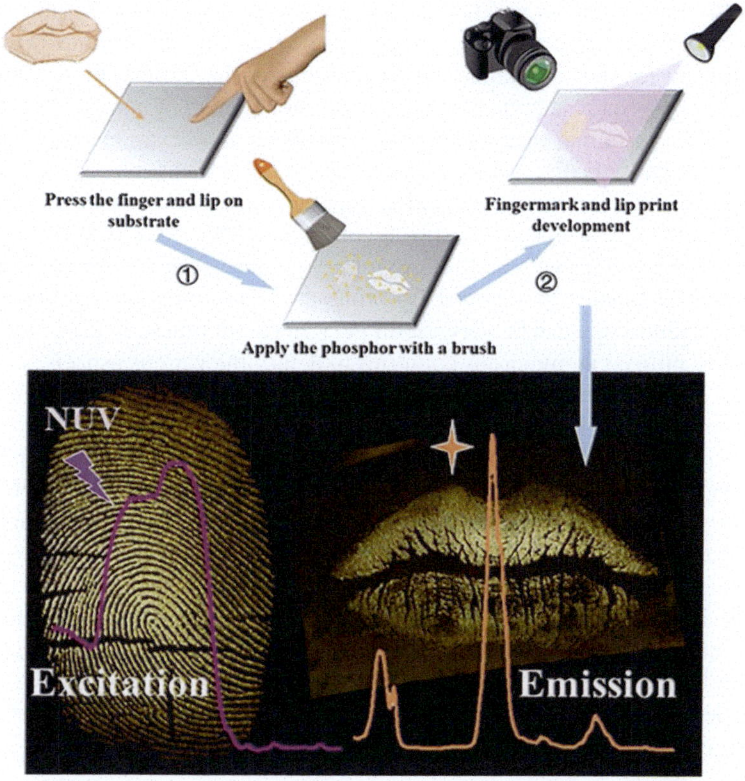

Fig. 4.2 MOF for lip and palm print detection. (Reprint with permission of Copyright- 2022, Elsevier, [45])

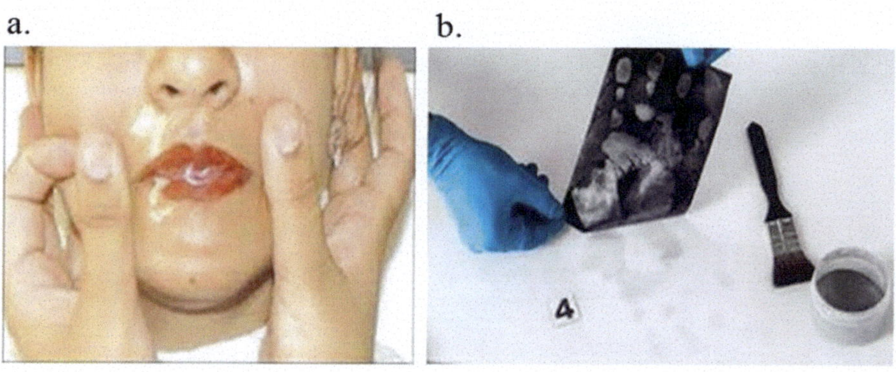

Fig. 4.3 (**a**) Lip print collection, (**b**) Palm print collection [50]

study of lip print is known as Cheiloscopy. This study focuses on the normal lines and cracks in the form of wrinkles and grooves, seen in the transitional area of the human lip between the inner labial mucosa and the outside skin. Many scientists are interested in lip-print research as a possible new component of human identification in both civil and criminal cases. Lip prints are utilized as an extra means of personal identification, and it is divergent even twins do not have identical.

Cheiloscopy is a forensic investigation approach that focuses on identifying people based on the traces left on their lips by other people [6, 33, 46]. When lip prints are found at a crime scene, cheiloscopy is very useful because they are so unique. For example, if lipstick blotches or lip prints are found on cups, glasses, or even envelopes, cheiloscopy can be very useful [47]. In recent times, lip prints have been increasingly utilized in numerous countries to determine various aspects such as an individual's gender, demographic characteristics, and even the place of residence across a country [48, 49].

On the other hand, palm prints are imprints of the friction ridges of the palm and fingers, and they can be used to identify a person. Palm prints are like lip prints and fingerprints are unique and distinct patterns found on the palm of individual, it can be classified into five types [51, 52] (Fig. 4.4). These patterns are formed by ridges, valleys, and creases on the surface of the palm and are unique to every individual for a lifetime. Palm prints are a valuable source for identification for the presence of the person at crime scene. The study of palm prints and fingerprints is dermatoglyphics. There are so many methods for taking the lip and palm impressions Fig. 4.2. In forensic detection, many of the existing treatments are used in forensic labs all over the world on a regular basis [53].

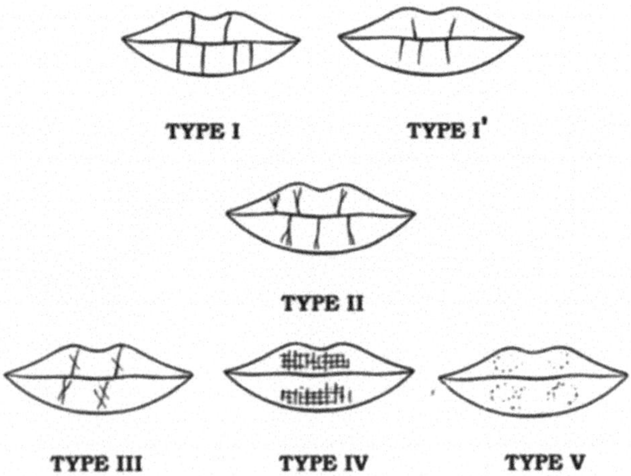

Fig. 4.4 Classification of lip prints. (Reprint with the permission [50])

4.2 MOF for Detection

4.2.1 MOF for Lip Print Detection

Lip prints are discovered in the same way as fingerprints. Lip prints may be found on glasses, cigarettes, envelopes, and other items on the investigation site. MOFs can be employed as sensing materials due to their high surface area, tunable pore sizes, and unique chemical properties. The MOF powder was put on the print with the help of a brush, and the shot was taken with a high-resolution camera under UV light. MOFs offer a promising avenue for the development of sensitive and selective sensors for detecting lip prints. By leveraging the unique properties of MOFs, researchers can create innovative and efficient methods for lip print analysis in forensic applications.

1. **Surface Area:** MOFs are purposefully designed to possess distinct surface characteristics that facilitate interactions with the residues present in lip prints. Lip prints are comprised of organic compounds, and specific MOFs possess the ability to selectively interact with these organic molecules via adsorption or other chemical interactions [54].
2. **Sensitivity and selectivity:** MOFs have the potential to be developed to be extremely sensitive and selective to chemical molecules that are found in lip prints. The basic characteristics of the material process can be modified throughout the MOF synthesis by selecting the suitable metal ions and ligands. This allows them to increase the sensitivity and selectivity of the material for lip print residue components [54].
3. **Fluorescence and colorimetric sensor:** Certain MOFs demonstrate changes in their fluorescence property or color when exposed to chemicals. Researchers can create sensors that exhibit color change or fluorescence [84] upon interaction with lip print residues by integrating fluorescent or colorimetric metal-organic frameworks MOFs. The variations can be readily identified and measured, hence offering a conspicuous indicator of the existence of lip impressions [55].
4. **Gas adsorption and desorption:** Volatile organic compounds or VOCs can be released from lip print of a person. Certain MOFs have been created to specifically adsorb and desorb certain gases or volatile organic compounds. Studies are able to passively detect and analyze lip print residues by monitoring the behavior of MOFs with regard to gas adsorption or desorption while the MOFs are present in the presence of lip prints [56].
5. **Portable devices:** It is possible to incorporate MOF-based lip print sensors into portable and handheld equipment, which enables on-site examination of lip prints in the context of forensic investigations. These tools are able to produce results quickly, which is beneficial to the work done by law enforcement organizations and forensic professionals.

4.2.1.1 Palm-Print Detection

The origins of palm print detection may be traced back to ancient civilizations, where handprints were utilized for many reasons, particularly identification. However, the development of modern scientific methodology started during the end of the nineteenth century with the groundbreaking contributions of Francis Galton [57]. Galton introduced a systematic categorization system for fingerprints, which encompassed palm prints as well [58–60]. This established the foundation for the systematic examination of biometric identification. Later contributions by scientists such as Henry Faulds and Edward Henry significantly expanded the area, particularly Henry's use of fingerprinting as a widely accepted technique for identifying criminals in the beginning of the twentieth century [61–63]. Technological progress in the twentieth century, including advancements in photography and computerized image analysis, improved palm print detection techniques. This led to the creation of automated fingerprint identification systems (AFIS) and played a role in the development of modern biometric authentication methods.

Palm print recognition is a biometric technique that identifies individuals by analyzing the unique pattern of their palm prints (Fig. 4.4). The palm print is a dependable means of human identification because of its unique patterns, which are not replicated in other individuals, including monozygotic twins [64]. Crucially, the characteristics of these ridges remain unchanged across time (Fig. 4.5). The ridge structures emerge during the 12th week of human embryonic development and reach completion by the 18th week [65]. The formation remains constant throughout life, except for its growth. The breakdown of the skin in the palm print area is the final stage after death. Palm print identification offers several advantages in comparison to other physical biometric traits. The elements encompassed in this list are low- resolution photography, little interference, uniform line characteristics, and the utilization of affordable capturing equipment [66].

Palm print detection involves a wide range of technical elements, all of which work together to precisely capture and analyze the distinct features of a palm of person. In order to begin this complex procedure, sophisticated imaging instruments such as optical scanners, cameras, or contactless sensors are used [68, 69]. These devices require high-resolution capability and accuracy to accurately reproduce the detailed ridge patterns, valleys, and minutiae points found on the surface of the palm. After capturing the palm print picture, it goes through a sequence of precise changes enabled by image processing algorithms. These algorithms provide essential functions like reducing noise, improving contrast, normalizing, and correcting geometric distortions in order to increase the quality of the captured picture. This ensures that the image is clear and accurate for further analysis.

A wide range of technical elements are included in palm print detection, all of which help to reliably record and evaluate the distinct features of a person's palm [66]. Advanced imaging tools like optical scanners, cameras, or contactless sensors are used to start this complex process [69–72]. Fig. 4.6. Pattern recognition algorithms are crucial in the following stage of palm print detection and identification. By utilizing the collected characteristics from the palm print picture, these algorithms aim to identify patterns and similarities between the taken palm print and the

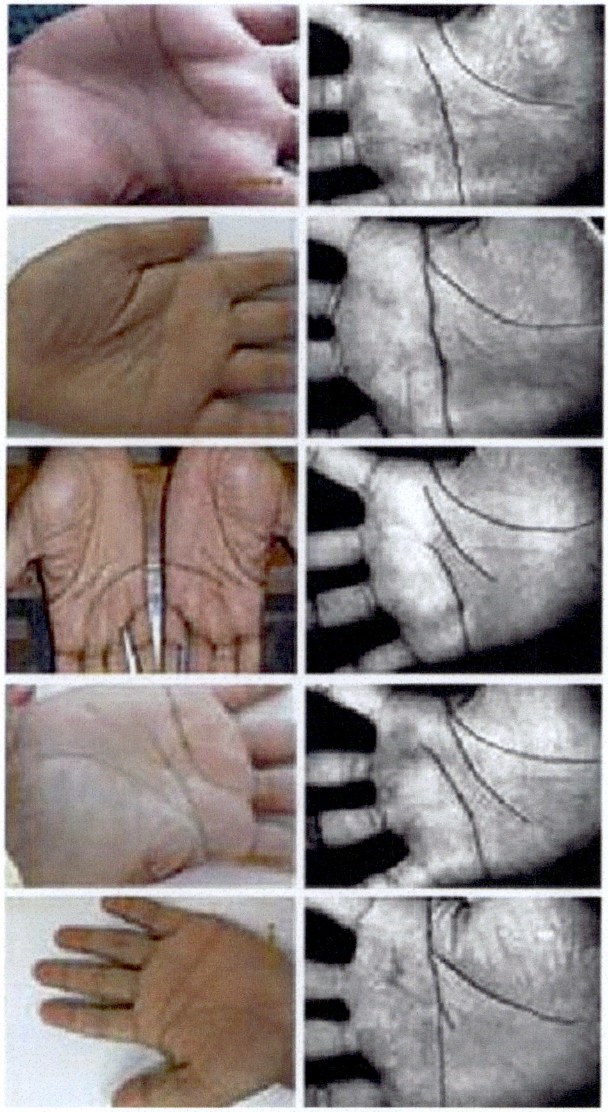

Fig. 4.5 Classification of palm prints

stored templates in a database [73]. Different matching strategies, including closest neighbor matching, minutiae-based matching, and correlation-based algorithms, are used to assess the level of similarity and identify possible matches. The act of comparing and analyzing is crucial for determining the identity linked to the palm print. In addition, the incorporation of machine learning and artificial intelligence methods has introduced a new period of complexity and effectiveness in palm print detecting systems [74]. These sophisticated approaches allow the system to adjust and improve its algorithms using gathered data and trends, hence increasing

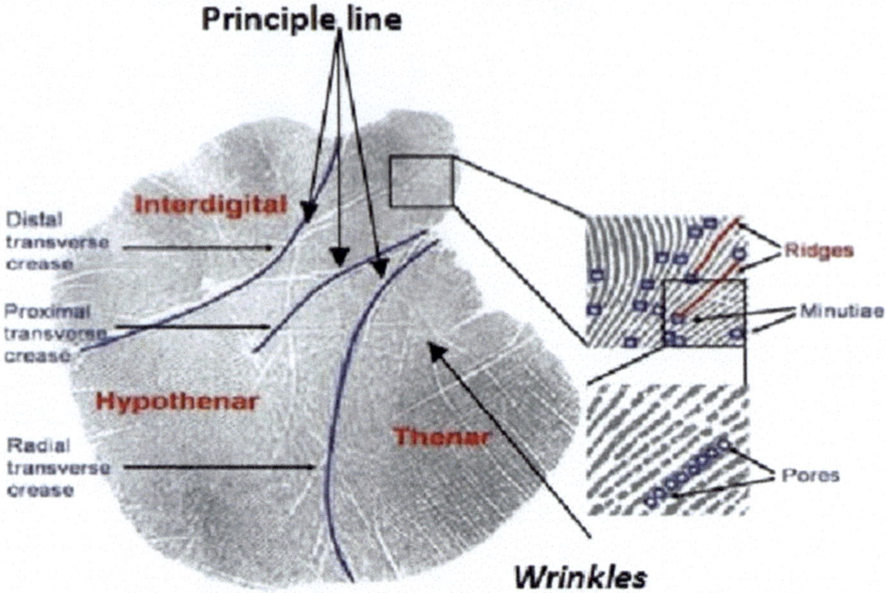

Fig. 4.6 Palm print image, principle line, ridges and valleys of palm [67]

accuracy, scalability, and flexibility. Modern palm print detection systems utilize extensive datasets and complex algorithms to attain exceptional levels of accuracy and dependability [74–76]. As a result, they have become indispensable instruments in several fields like law enforcement, border control, and access control. At present, the majority of palm print biometrics systems employ a scanning device or CCD camera as the input sensor [64]. Users are required to physically contact the sensor for their hand photos to be captured. The issue of sanitation is of paramount significance, particularly in public settings such as hospitals. Individuals are apprehensive about placing their fingers or palms on a shared sensor that has been touched by several other individuals. This issue is notably intensified in several Asian nations during the peak of the SARS pandemic (Fig. 4.7).

Furthermore, latent palm prints that persist on the surface might be replicated for illicit purposes. In addition, if not utilized correctly, particularly in severe, dusty, and outdoor settings, the surface will become readily polluted. Furthermore, many traditionalist countries may exhibit reluctance to place their hands on the sensor following contact by an individual of the opposing gender. Thus, there is an urgent requirement for biometric technology that is sufficiently adaptable to record the users hand photographs without necessitating physical contact with the sensor platform [78, 79].

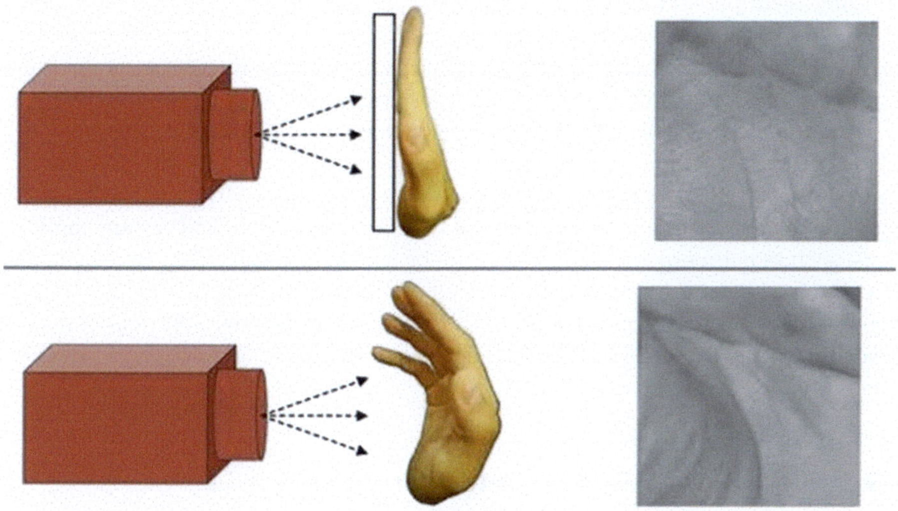

Fig. 4.7 Images of contact less palm print detection [77]

4.2.2 MOF for Palm Print Detection

The utilization of MOFs in the fields of forensic science and biometrics has garnered significant attention in scientific circles. As discussed earlier, due to unique properties of MOF, it is used for detection of palm print, which can be tailored for specific application.

This is primarily attributed to the distinctive characteristics exhibited by MOFs, which may be customized to suit certain applications. The following are potential ways in which MOFs could be applied in this context:

1. **Adsorption and analysis:** MOFs possess the ability to be modified to effectively capture VOCs or molecules that emanate from the skin and are seen in palm prints. By strategically engineering MOFs with a pronounced affinity for these specific molecules, they have the possibility to be used in the isolation and concentration of minute constituents found in palm prints, hence facilitating subsequent analytical procedures [56].
2. **Colorimetric or fluorescent Sensor:** When they are subjected to certain compounds, certain MOFs undergo color or fluorescence changes. MOFs might be designed by researchers to interact with the chemicals in palm prints, resulting in a noticeable color change or fluorescence in the print. This modification might be able to be identified and utilized in palm print analysis [55].
3. **Gas sensing devices:** MOFs have high sensitivity to particular gases or VOCs could be included into sensors that are used to detect and analyze gases that are generated by the skin. These sensors could include components of palm prints. For the purposes of identification or analysis, changes in the adsorption or desorption of gases might be monitored [80].

4. **Long-term preservation:** MOFs have the potential to be employed in the preservation of palm prints through the process of encapsulation for their long-term storage and protection. This could shield the prints from the effects of environmental conditions and processes of deterioration, making it possible to store them for a longer period of time and conduct study on them [81].

In summary it is essential to notice that the production of MOFs and their application in the field of forensics is a topic of continuing research, and certain applications, such as palm print detection, may not be as widely used as they may be in the near future. If you are interested in the most recent advancements in this field, I suggest that you look through the most recent scientific literature and research papers for updates on the use of MOFs in forensic applications, such as the identification of palm prints.

4.2.3 Compounds Used for Detection

Just like fingerprints, lip prints were analyzed by putting a lipstick or moisturizer on lips then gently pressing the lips against the surface to develop the print on the surface foil paper or on steel surface, mobile screen, black colored paper. Studies are still going on for the production of MOF compounds that have been used for lip print and palm print detection. Initially Dhanalakshmi et al. synthesized the powdered compound $BaTiO3:Dy^{3+}$ for the detection of lip print.[82] Furthermore, Sharma et al. synthesized a nanophosphor $CaAl_{2O4:\ Eu}{}^{2+}$, Dy^{3+} to develop fingerprint and lip print [83]. Li et al. synthesized a yellow phosphor $LaNb_{2VO9:0.10Dy}{}^{3+}$ for detection of lip print for forensic investigation [45]. The detection was proceeded when the formed nano powder, nanophosphor, and phosphor sprinkled by the using a brush and analyzed under UV light at 275 nm on the prints taken on surfaces from the sites.

4.3 Conclusion

MOF has great potential in the field of forensic science for the detection and analysis of lip prints and palm prints. These distinctive properties such as high surface area, tunable porosity, and selective adsorption capabilities make MOF a valuable material for developing sensitive detection methods. Interestingly, MOF can be engineered in a way to interact selectively with the organic compounds that are present in Lip prints of human beings. Similarly, MOF can be designed with high affinity, that can be designed to adsorb VOCs utilized to concentrate and isolate trace elements that are present in palm prints of human beings. MOFs can also protect lip and palm prints from decomposition and environmental causes. It is important to note that to optimize the use of MOFs for forensics, continuing research and improvements must be conducted. Further investigation of MOF-based detection technologies could revolutionize forensic analysis by improving the accuracy, reliability, and efficiency of lip and palm print identification.

References

1. Yadav N, Mudgal D, Mishra A, Shukla S, Malik T, Mishra V. Harnessing fluorescent carbon quantum dots from natural resource for advancing sweat latent fingerprint recognition with machine learning algorithms for enhanced human identification. PLoS One. 2024;19(1 January):e0296270. https://doi.org/10.1371/journal.pone.0296270.
2. Augustine J, Barpande SR, Tupkari JV. Cheiloscopy as an adjunct to forensic identification: a study of 600 individuals. J Forensic Odonto-Stomatol JFOS. 2008;26(2):44–52.
3. Limson KS, Julian R. Computerized recording of the palatal rugae pattern and an evaluation of its application in forensic identification. J Forensic Odonto- Stomatol JFOS. 2004;22(1):1–4.
4. Castello A, Verdu F. Development of latent lip prints on multicoloured surfaces, a problem resolved using fluorescent dyes. Indian Cong Forensic Med Toxicol. 2006
5. Prabhu RV, Dinkar A, Prabhu V. Digital method for lip print analysis: a new approach. J Forensic Dent Sci. 2013;5(2):96.
6. Vamsi L, Reddy K. Lip prints: an overview in forensic dentistry. J Adv Dent Res. 2011; https://doi.org/10.1177/2229411220110104.
7. Bhattacharjee R, Kar AK. Cheiloscopy: a crucial technique in forensics for personal identification and its admissibility in the Court of Justice. Morphologie. 2024;108(360):100701.
8. Šimović M, Pavušk I, Muhasilović S, Vodanović M. Morphologic patterns of lip prints in a sample of Croatian population. Acta Stomatol Croat. 2016;50(2):122.
9. Ramakrishnan K, Sharma S, Sreeja C, Pratima DB, Aesha I, Vijayabanu B. Sex determination in forensic odontology: a review. J Pharm Bioallied Sci. 2015;7(Suppl 2):S398–402.
10. Kasprzak J. Possibilities of cheiloscopy. Forensic Sci Int. 1990;46(1–2):145–51.
11. Saraswathi T, Mishra G, Ranganathan K. Study of lip prints. J Forensic Dent Sci. 2009:28–31.
12. Sunday OG, Stephen AA, Wokpeogu PC, Nmereni AP, Nwolim PJ, Owabhel RF. Cheiloscopy among the Igbo students in Madonna University Elele, Rivers state, Nigeria. Saudi J Biomed Res. 2018;3(3):129–35.
13. Sivapathasundharam B, Prakash PA, Sivakumar G. Lip prints (cheiloscopy). Indian J Dent Res. 2001;12(4):234–7.
14. Bharathi S, Thenmozhi MS. Cheiloscopy-lip print, an determination of sex and individual. J Pharm Sci Res. 2015;7(6):330.
15. Prabhu RV, Dinkar AD, Prabhu VD. Collection of lip prints as a forensic evidence at the crime scene—an insight. J Oral Health Res. 2010;1(4).
16. Ball J. The current status of lip prints and their use for identification. J. Forensic Odonto-Stomatol JFOS. 2002;20(2):43–6.
17. Verma Y, et al. A study of lip prints and its reliability as a forensic tool. Natl J Maxillofac Surg. 2015;6(1):25–30.
18. Dwivedi N, Agarwal A, Kashyap B, Raj V, Chandra S. Latent lip print development and its role in suspect identification. J Forensic Dent Sci. 2013;5(1):22.
19. Utsuno H, Kanoh T, Tadokoro O, Inoue K. Preliminary study of post mortem identification using lip prints. Forensic Sci Int. 2005;149(2–3):129–32.
20. Jaber M, Varma SR, Hassan M, Hadi K, Khamis A. Assessment of lip print (cheiloscopy) patterns and its use for personal identification and crime investigation: a systematic review and meta-analysis. J Int Dent Med Res. 2023;16(1):331–9.
21. Abedi M, Afoakwah C, Bonsu DNOM. Lip print enhancement. Forensic Sci Res. 2022;7(1):24–8.
22. Abebe B, Murthy HCA, Zereffa EA, Dessie Y. Latent fingerprint enhancement techniques: a review. J Chem Rev. 2020;2(1):40–56.
23. Singh S, Singh A, Sah K, Singh M. Latent lip print: a comparative study of developers. J Oral Med Oral Surgery Oral Pathol Oral Radiol. 2023;7(4):216–21.
24. Asli K. Lip print analysis in Malaysian Chinese population (klang valley): lipstick- cellophane tape technique. J Sains Kesihat Malaysia. 2020;18(2):31–8.

25. Kapoor N, Badiye A. A study of distribution, sex differences and stability of lip print patterns in an Indian population. Saudi J Biol Sci. 2017;24(6):1149–54.
26. Welgemoed M. Lip prints: the underestimated identifiers in the combat against crime. Obiter. 2019;40(2):281–306.
27. Mahmoud NF, Afify MM, Elbendary RN, Shokry DA. Variations in morphological patterns of lip prints as evidence in racial and sexual discrimination. J Forensic Res. 2020;11(1):1–7.
28. Kesarwani P, Choudhary A. Correlation of lip print with blood group in forensic science. J Oral Maxillofac Pathol. 2021;25(1):206.
29. Chopra M. The personal identification by means of cheiloscopy among people of North India. Indian J Forensic Med Toxicol. 2020;14(4).
30. Castelló A, Alvarez-Seguí M, Verdú F. Luminous lip-prints as criminal evidence. Forensic Sci Int. 2005;155(2–3):185–7.
31. Singh NN, Brave VR, Khanna S. Natural dyes versus lysochrome dyes in cheiloscopy: a comparative evaluation. J Forensic Dent Sci. 2010;2(1):11.
32. Dolly A, Rodrigues C, Bankur R, Gopinathan PA, Sharma R, Doddamani A. Evaluation of efficacy of three different materials used in cheiloscopy–a comparative study. J Clin Diagn Res JCDR. 2016;10(10):ZC67.
33. Fonseca GM, Ortíz-Contreras J, Ramírez-Lagos C, López-Lázaro S. Lip print identification: current perspectives. J Forensic Leg Med. 2019;65(October):32–8. https://doi.org/10.1016/j.jflm.2019.04.009.
34. Agarwal A, Dwivedi N, Khare P. Latent lip print: a new possibility. Indian J Forensic Sci. 2013;6(1).
35. Devi A, Astekar M, Kumar V, Kaur P, Singh N, Sidhu GK. The study of inheritance analysis and evaluation of lip prints in individuals. J Forensic Dent Sci. 2015;7(1):49.
36. Sharma P, Sharma N, Aggarwal P. Latent lip prints: how far a link to the suspect? Eur J Forensic Sci. 2015;2(4):11.
37. Seguí MA, Feucht MM, Ponce AC, Pascual FAV. Persistent lipsticks and their lip prints: new hidden evidence at the crime scene. Forensic Sci Int. 2000;112(1):41–7.
38. U. S. F. B. of Investigation, et al. Processing guide for developing latent prints. Federal Bureau of Investigation; 2001.
39. Castello A, Alvarez M, Miquel M, Verdu F. Long-lasting lipsticks and latent prints. Forensic Sci Commun. 2002;4(2).
40. Wang M, Li M, Yu A, Zhu Y, Yang M, Mao C. Fluorescent nanomaterials for the development of latent fingerprints in forensic sciences. Adv Funct Mater. 2017;27(14):1606243.
41. Yusuf VF, Malek NI, Kailasa SK. Review on metal–organic framework classification, synthetic approaches, and influencing factors: applications in energy, drug delivery, and wastewater treatment. ACS Omega. 2022;7(49):44507–31.
42. Moret S, et al. Metal-organic frameworks for fingermark detection—a feasibility study. Forensic Sci Int. 2018;291:83–93. https://doi.org/10.1016/j.forsciint.2018.08.005.
43. Sharma V, Das A, Kumar V, Kumar V, Verma K, Swart HC. Combustion synthesis and characterization of blue long lasting phosphor CaAl 2 O 4: Eu 2 + , Dy 3 + and its novel application in latent fingerprint and lip mark detection. Phys B Phys Condens Matter. 2018;535(July):149–56. https://doi.org/10.1016/j.physb.2017.07.019.
44. Dhanalakshmi M, Nagabhushana H, Darshan GP, Basavaraj RB. Sonochemically assisted hollow / solid BaTiO 3: Dy 3 þ microspheres and their applications in effective detection of latent fingerprints and lip prints. J Sci Adv Mater Dev. 2017;2(1):22–33. https://doi.org/10.1016/j.jsamd.2017.02.004.
45. Li S, et al. Synthesis and characterization of wide band excited yellow phosphor LaNb2VO9:Dy3+ and application in potential fingerprint and lip print detection. J Lumin. 2022;244(August):118681. https://doi.org/10.1016/j.jlumin.2021.118681.
46. Dineshshankar J, Ganapathi N, Yoithapprabhunath TR, Maheswaran T, Kumar MS, Aravindhan R. Lip prints: role in forensic odontology. J Pharm Bioallied Sci. 2013;5(SUPPL):1. https://doi.org/10.4103/0975-7406.113305.

47. Wrobel K, Doroz R, Porwik P, Bernas M. Personal identification utilizing lip print furrow based patterns. A new approach. Pattern Recogn. 2018;81:585–600. https://doi.org/10.1016/J.PATCOG.2018.04.030.
48. Mishra G, Ranganathan K, Saraswathi T. Study of lip prints. J Forensic Dent Sci. 2009;1(1):28. https://doi.org/10.4103/0974-2948.50885.
49. Wrobel K, Doroz R, Porwik P, Naruniec J, Kowalski M. Using a probabilistic neural network for lip-based biometric verification. Eng Appl Artif Intell. 2017;64(April):112–27. https://doi.org/10.1016/j.engappai.2017.06.003.
50. Gardner ES. No 主観的健康感を中心とした在宅高齢者における 健康関連指標に関する共分散構造分析Title, 1960;(5):63–65.
51. Rekha VR, Sunil S, Rathy R. Heritability and correlation of lip prints and palmprints in South Kerala population. Oral Maxillofac Pathol J. 2015;6(1):544–9.
52. Police BA. Palm print identification in forensic sciences, 2022:1–9.
53. Selbes B, Elihoş A. Deep learning based latent palmprint recognition. In: 2023 31st signal processing and communications applications conference (SIU), vol. 2023. IEEE. p. 1–4.
54. Mousavi SM, et al. Innovative metal-organic frameworks for targeted oral cancer therapy: a review. Materials (Basel). 2023;16(13):4685.
55. Pal TK. Metal–organic framework (MOF)-based fluorescence 'turn-on' sensors. Mater Chem Front. 2023;7(3):405–41.
56. Xie Y, Lyu S, Zhang Y, Cai C. Adsorption and degradation of volatile organic compounds by metal–organic frameworks (MOFs): a review. Materials (Basel). 2022;15(21):7727.
57. Cole S. What counts for identity? The historical origins of the methodology of latent fingerprint identification. Sci Context. 1999;12(1):139–72.
58. Berry J, Stoney DA. The history and development of fingerprinting. Adv Fingerpr Technol. 2001;2:13–52.
59. Hawthorne M, Plotkin S, Douglas B-A. Fingerprints: analysis and understanding the science. CRC Press; 2021.
60. Basu S. History of fingerprint in India: unraveling a truth for scientific certainty, 2020.
61. Cole SA. Manufacturing identity: a history of criminal identification techniques from photography through fingerprinting. Cornell University; 1998.
62. Hoover JE. The role of identification in law enforcement: an historical adventure. John's L Rev. 1971;46:613.
63. Ferguson R. Identification practices in twentieth-century fiction. Oxford University Press; 2021.
64. Michael GKO, Connie T, Teoh ABJ. Touch-less palm print biometrics: novel design and implementation. Image Vis Comput. 2008;26(12):1551–60.
65. Cummins H, Midlo C. Finger prints, palms and soles: an introduction to dermatoglyphics, vol. 319. Dover Publications New York; 1961.
66. Amrouni N, Benzaoui A, Zeroual A. Palmprint recognition: extensive exploration of databases, methodologies, comparative assessment, and future directions. Appl Sci. 2023;14(1):153.
67. Ali MMH, Yannawar P, Gaikwad AT. Study of edge detection methods based on palmprint lines. In: 2016 international conference on electrical, electronics, and optimization techniques (ICEEOT). IEEE; 2016. p. 1344–50.
68. Mahalakshmi BS, Sheela SV. A review on palmprint authentication system using various feature extraction and classification techniques. 7:36–43.
69. Alausa DWS, et al. Contactless palmprint recognition system: a survey. IEEE Access. 2022;10:132483–505.
70. Das P, Meshram S. A review paper based on palmprint recognition system. Int J Eng Res Technol. 2014;3(2).
71. Al-Taie SAM, Khaleel BI. Palm print recognition using intelligent techniques: a review. J Ilm Tek Elektro Komput dan Inform. 2023;9(1):156–64.
72. Michael GKO, Connie T, Teoh ABJ. A contactless biometric system using palm print and palm vein features. Adv Biometric Technol. 2011:155–77.
73. Mehmood AB, Taj IA, Ghafoor M. Palmprint enhancement network (PEN) for robust identification. Multimed Tools Appl. 2024;83(5):14449–76.

74. Sarker IH. Machine learning: algorithms, real-world applications and research directions. SN Comput Sci. 2021;2(3):160.
75. Trabelsi S, Samai D, Dornaika F, Benlamoudi A, Bensid K, Taleb-Ahmed A. Efficient palmprint biometric identification systems using deep learning and feature selection methods. Neural Comput Appl. 2022;34(14):12119–41.
76. Jaswal G, Poonia RC. Selection of optimized features for fusion of palm print and finger knuckle-based person authentication. Expert Syst. 2021;38(1):e12523.
77. Stoimchev M, Ivanovska M, Štruc V. Learning to combine local and global image information for contactless palmprint recognition. Sensors. 2021;22(1):73.
78. Le HV, Kosch T, Bader P, Mayer S, Henze N. PalmTouch: using the palm as an additional input modality on commodity smartphones. In: Proceedings of the 2018 CHI conference on human factors in computing systems; 2018. p. 1–13.
79. Chuen Lee L, Izzati Bohari N. Forensic fingerprint analysis using self- organizing maps, classification and regression trees and Naïve Bayes methods. Int J Comput Digit Syst. 2021.
80. Zhang R, Lu L, Chang Y, Liu M. Gas sensing based on metal-organic frameworks: concepts, functions, and developments. J Hazard Mater. 2022;429:128321.
81. Saeb MR, Rabiee N, Mozafari M, Mostafavi E. Metal-organic frameworks (MOFs)-based nanomaterials for drug delivery. Materials (Basel). 2021;14(13):3652.
82. Dhanalakshmi M, Nagabhushana H, Darshan GP, Basavaraj RB, Daruka Prasad B. Sonochemically assisted hollow/solid BaTiO3:Dy3+ microspheres and their applications in effective detection of latent fingerprints and lip prints. J Sci Adv Mater Dev. 2017;2(1):22–33. https://doi.org/10.1016/j.jsamd.2017.02.004.
83. Sharma V, Das A, Kumar V, Kumar V, Verma K, Swart HC. Combustion synthesis and characterization of blue long lasting phosphor CaAl2O4: Eu2+, Dy3+ and its novel application in latent fingerprint and lip mark detection. Phys B Condens Matter. 2018;535:149–56. https://doi.org/10.1016/j.physb.2017.07.019.
84. Guliani E. Taneja A, Ranjan KR, Mishra V. Luminous insights: Exploring organic fluorescent "Turn-On" chemosensors for metal-ion (Cu+ 2, Al+ 3, Zn+ 2, Fe+ 3) detection. J Fluoresc. 2024;34(5):1965–2001.
85. Dwivedi B, Mishra PK, Mishra V. Chiral metal-organic frameworks for asymmetrical catalysis. In Metal-organic frameworks (MOFs) as catalysts, Singapore: Springer Nature Singapore, 2022;493–513.

Metal-Organic Frameworks for Detection of Pesticides in Forensic Samples

5

Abstract

In forensic science, the detection and analysis of pesticides in various samples are of paramount importance for investigating cases related to poisoning, environmental contamination, and agricultural crimes. Metal-organic frameworks (MOFs) have emerged as promising materials in various fields due to their tunable structures, high surface areas, and customizable functionalities. This chapter provides an overview of the utilization of MOFs as sensing platforms for the detection of pesticides in forensic samples. It covers the design principles, fabrication methods, and sensing mechanisms of MOF-based sensors. Additionally, the chapter discusses the application of MOF-based sensors in the detection of specific pesticides, highlighting their sensitivity, selectivity, and potential for on-site analysis. Furthermore, challenges and future perspectives in the development of MOF-based sensors for forensic applications are addressed, aiming to inspire further research and innovation in this rapidly evolving field.

5.1 Introduction

Pesticides are the chemicals extensively used to prevent, destroy, or mitigate pest in the field of agriculture [1]; nevertheless, the residues they leave behind have the potential to inflict significant harm on ecological integrity. Furthermore, these residues may have detrimental effects on human health and overall quality of life since they can be transmitted down the food chain. Therefore, the expeditious and discerning identification of pesticides has emerged as a paramount problem pertaining to human well-being and the safeguarding of the environment [2]. There is a broad spectrum of toxicity that can be presented by pesticides. Because of its potential toxicity, pesticides are frequently employed as poison for a variety of reasons, including as a method of suicide by farmers, accidental poisoning, homicide, and

V. Mishra, A. Mishra, *Metal-Organic Frameworks as Forensic Detectors*,
https://doi.org/10.1007/978-981-96-2147-7_5

animal cruelty. Viscera, food samples, and liquid samples are all sent to the forensic laboratory to be tested for the presence of pesticides [1]. Conventional methods such as LC-MS, GC-MS, or FTIR are only some of the many methods that may be utilized in the process of detecting pesticides. On the other hand, these methods have several drawbacks, such as the fact that the operation takes a long amount of time, is expensive, and requires complex apparatus [3]. In recent years, several Metal-Organic Frameworks (MOFs) have emerged as a promising sensory system for the highly sensitive detection of pesticides. The distinct porous structures and tunable properties of MOFs make it possible for them to recognize pesticides in an effective and selective manner, and they may offer potential solutions for addressing environmental and food safety concerns [4]. Metal-organic frameworks (MOFs) are a kind of crystalline material that exhibit a hybrid structure consisting of both organic and inorganic components [5]. These components, namely organic ligands, and metal ions, are interconnected in three-dimensional space, giving rise to frameworks that may be one-dimensional (1-D), two-dimensional (2-D), or three-dimensional (3-D) in nature [6] (Fig. 5.1).

Metal-organic frameworks (MOFs) provide many advantageous characteristics that confer them a competitive edge over traditional methodologies. These include notable selectivity, a remarkable level of flexibility, and an expedited reaction rate [4]. Ability of various MOFs to detect pesticides and their limit of detection are summarized in Table 5.1.

The table presents a diverse range of Metal-Organic Frameworks (MOFs) used for pesticide sensing, showcasing various sensing categories and their corresponding detection limits. Notably, Zr-based MOFs exhibit exceptional sensitivity in

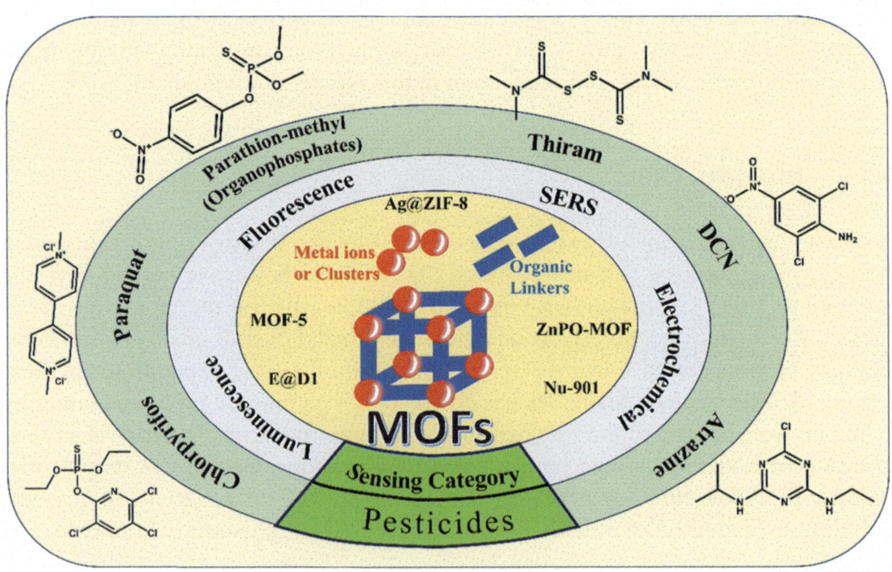

Fig. 5.1 MOFs for pesticide detection

Table 5.1 Ability of various MOFs to detect pesticides and limit of detection

S. No	MOFs	Pesticide	Sensing category	Detection limit	Ref.
1.	Zr-based MOFs	Malathion Parathion-methyl	Luminescence	4.9×10^{-15} M 0.438 nM	[7]
2.	MOF-5	Nitro organophosphates	Luminescence	5 ppb	[8]
3.	Ag@ZIF-8	Thiram	SERS	0.1 μM	[9]
4.	AuNP/MOF-199	Acetamiprid	SERS	0.02 μM	[10]
5.	AuNP/UiO-66	Acetamiprid	SERS	0.009 μM	[10]
6.	Fe_3O_4-Au@ MIL-100(Fe)	Thiram	SERS	15 nM	[11]
7.	$[Cd(atc)(H_2O)_2]_n$	Parathion	Electrochemical	0.14 ng/ml	[12]
8.	Eu^{3+}@1	p-nitrophenol 3-methyl-4-nitrophenol	Luminescence (competitive absorption)	0.36 μg/ml 0.41 μg/ml	[13]
9.	Mn/Fe-MIL (53)	Methyl parathion Chlorpyrifos	Luminescence	2.8 nM 0.95 nM	[14]
10.	AuNCs@ZIF-8	Organophosphates.	Luminescence	0.3 μg/l	[15]
11.	$[Cd(NH_2\text{-bdc})(azp)]\cdot DMF$	Glyphosate	Structural collapse of LMOFs	25 nM	[16]
12.	ZnPO-MOF	Parathion-methyl	Luminescence	0.456 nM (0.12 g/kg	[17]
13.	$[Tb(L)_2NO_3]_n$	Glyphosate	Luminescence (PET)	0.0144 μM	[18]
14.	$Tb(tftpa)_{1.5}(2,2'\text{-}bpy)(H_2O)$	Chlorpyrifos	Fluorescence	0.14 ppb	[19]
15.	$[Zn_2(cptpy)(btc)(H_2O)]_n$	Paraquat	Luminescence	9.73 μM	[20]
16.	$[Cd(tptc)_{0.5}(bpy)]_n$	DCN	Luminescence	638 ppb	[21]
17.	NU-901	Paraquat	Luminescence (PET)	2.0 nM	[22]
18.	$[Zn_3(DDB)(DPE)]\cdot H_2O$	2,6-Dichloro-4-nitroamine (2,6-Dich-4-NA)	Fluorescence	0.27 μM	[23]
19.	$Cu_2(BTC)_2$ @SiO_2	Atrazine	Electrochemical	0.01 nM	[24]
20.	$[Ag(CIP^-)]$	DCN	Fluorescence	0.17 μM	[25]

luminescence-based detection, with a detection limit as low as 4.9×10^{-15} M for Malathion. Several other MOFs, such as Ag@ZIF-8, AuNP/MOF-199, and Mn/Fe-MIL (53), offer impressive capabilities in Surface-Enhanced Raman Spectroscopy (SERS) for detecting pesticides like Thiram and Acetamiprid. Furthermore, electrochemical sensing with MOFs, as demonstrated by $[Cd(atc)(H_2O)_2]n$, can detect Parathion at an astonishingly low concentration of 0.14 ng/mL. The table reflects the versatility and potential of MOFs in addressing the critical need for highly sensitive pesticide detection methods. In their study, Kumar et al. investigated the impact of several nitro organophosphates on the luminescent properties of MOF-5, a metal-organic framework [8]. The synthesis of MOF-5 was achieved by the reaction between zinc nitrate and terephthalic acid in a diethylformamide (DEF) solvent. Significantly, the synthesized metal-organic framework (MOF) exhibited noteworthy fluorescence properties, as shown by its excitation wavelength of 330 nm and emission wavelength of 493 nm. The study has also shown that the zinc-based metal-organic framework (MOF) has significant promise in the detection of organophosphate pesticides (OPs) containing nitro groups, such as parathion, methyl parathion, paraoxon, and fenitrothion. Remarkably, the use of the MOF-assisted methodology enabled the identification of these four organophosphates (OPs) separately, at concentrations spanning from 5 to 600 parts per billion (ppb), while achieving a detection threshold of 5 ppb. Despite the inclusion of these additional operational procedures, the approach maintained its characteristic level of accuracy. Significantly, the sensing technique that was identified exhibited selectivity towards nitro OPs, while being unaffected by other OPs such as malathion, dichlorvos, and monocrotophos [8]. Cai et al. explored AuNCs@ZIF-8 for detection of OPs with quantum yield of 4.63% [15]. Luo et al. constructed Mn/ Fe-MIL (53) MOF which can quantitively analyze chlorpyrifos and methyl parathion in concentration range of 5–50 nM and 10–20 nM respectively [26]. LOD for methyl parathion and chlorpyrifos were 0.95 nM and 2.8 nM respectively as shown in Fig. 5.2.

Tao et al. in 2017 discussed a new metal-organic framework (MOF) denoted as $[Zn_2(bpdc)_2(BPyTPE)]$ [2]. This MOF exhibits a unique structure characterized by pillared-layered entanglement and luminous properties. The compound (*E*)-1,2-diphenyl-1,2-bis(4-(pyridin-4-yl) phenyl)ethene, also known as BPyTPE, was synthesized with a ligand based on tetraphenylethene (TPE) with significant aggregation-induced emission (AIE) characteristics. BPyTPE was then used as a linker to produce a unique pillared-layered metal-organic framework (LMOF). The synthesis of complex 1 involves the reaction of BPyTPE with $Zn(NO_3)_2{\bullet}6H_2O$ and biphenyl-4,4′-dicarboxylic acid (H_2bpdc) in a solvent mixture of

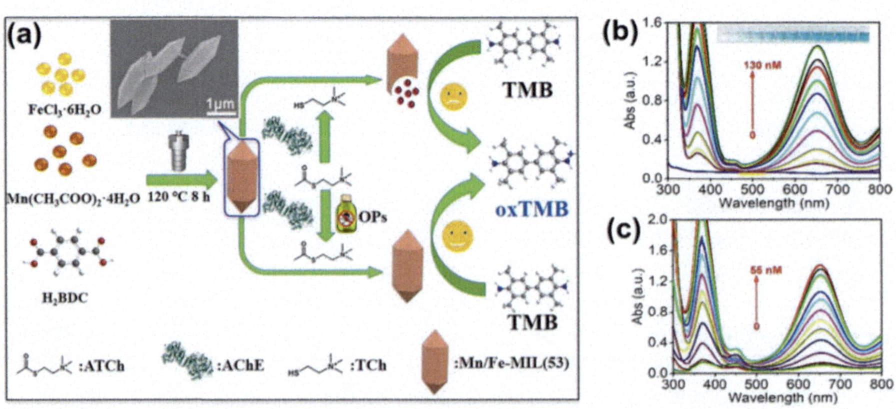

Fig. 5.2 (**a**) Sensing technique of Mn/Fe-MIL (53) for organophosphate detection. (**b**) LOD for chlorpyrifos (**c**) LOD for methyl parathion. (Adapted from [14])

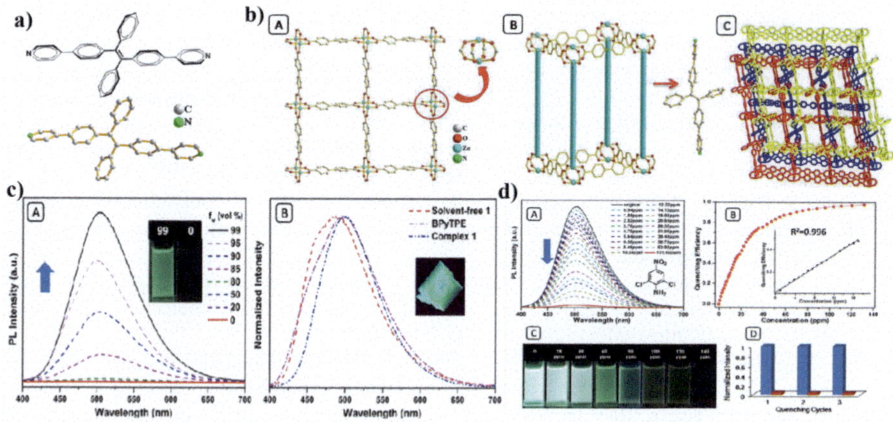

Fig. 5.3 (**a**) Molecular and single-crystal structures of AIE-active BPyTPE. (**b**, **d**) Schematic representation for a highly luminescent [$Zn_2(bpdc)_2(BPyTPE)$] MOF and its application for 2,6-dichloro-4-nitroaniline pesticide detecting. (**c**) (A) Photoluminescence (PL) spectra of BPyTPE in THF/water mixtures with various water fractions (*f*w). (Reproduced with permission from Ref. [2] copyright 2017 RSC)

N,N-dimethylacetamide and N,N-dimethylformamide. This reaction results in the formation of colorless crystals with a cubic morphology. The solvent-free compound 1 demonstrates a robust emission in the blue-green region, characterized by an exceptional fluorescence quantum yield of 99%. Additionally, it offers a straightforward and reversible approach for the sensitive and quantitative detection of trace amounts of 2,6-dichloro-4-nitroaniline, a pesticide. The ligand BPyTPE, which is based on TPE (tetraphenylethene). This conformation is further verified through analysis of the single-crystal structure, as seen in Fig. 5.3.

References

1. Mishra A, Srivastava A, Kanojia R, Mishra V. Improved isolation and detection approach of Chlorpyrifos pesticide from urine and viscera samples. J Forensic Med Toxicol. 2022;39(1):45–50.
2. Tao C-L, et al. A highly luminescent entangled metal–organic framework based on pyridine-substituted tetraphenylethene for efficient pesticide detection. Chem Commun. 2017;53(72):9975–8.
3. Zhao D, et al. Recent progress in metal-organic framework based fluorescent sensors for hazardous materials detection. Molecules. 2022;27(7):2226. https://doi.org/10.3390/molecules27072226.
4. Xue Y, Peng Y, Geng Z, Wang Y, Ung COL, Hu H. Metal–organic frameworks (MOFs) based analytical techniques for food safety evaluation. eFood. 2021;2(1):1. https://doi.org/10.2991/efood.k.210209.001.
5. Dwivedi B, Mishra PK, Mishra V. In: Gulati S, editor. Chiral metal-organic frameworks for asymmetrical catalysis BT - metal-organic frameworks (MOFs) as catalysts. Singapore: Springer; 2022. p. 493–513. https://doi.org/10.1007/978-981-16-7959-9_19.

6. Aquino A, Alves Ferreira J, Navickiene S, Wanderley KA, de Sá GF, Júnior SA. Investigating the potential of metal-organic framework material as an adsorbent for matrix solid-phase dispersion extraction of pesticides during analysis of dehydrated Hyptis pectinata medicinal plant by GC/MS. J AOAC Int. 2012;95(5):1338–42. https://doi.org/10.5740/jaoacint.SGE_Aquino.
7. Ma L, et al. Nanocomposites of Pt nanoparticles anchored on UiO66-NH2 as carriers to construct acetylcholinesterase biosensors for organophosphorus pesticide detection. Electrochim Acta. 2019;318:525–33. https://doi.org/10.1016/j.electacta.2019.06.110.
8. Kumar P, Paul AK, Deep A. Sensitive chemosensing of nitro group containing organophosphate pesticides with MOF-5. Microporous Mesoporous Mater. 2014;195:60–6. https://doi.org/10.1016/j.micromeso.2014.04.017.
9. Li Q, Gong S, Zhang H, Huang F, Zhang L, Li S. Tailored necklace-like Ag@ZIF-8 core/shell heterostructure nanowires for high-performance plasmonic SERS detection. Chem Eng J. 2019;371:26–33. https://doi.org/10.1016/J.CEJ.2019.03.236.
10. Cao X, et al. SERS-active metal–organic frameworks with embedded gold nanoparticles. Analyst. 2017;142(14):2640–7. https://doi.org/10.1039/C7AN00534B.
11. Lai H, Shang W, Yun Y, Chen D, Wu L, Xu F. Uniform arrangement of gold nanoparticles on magnetic core particles with a metal-organic framework shell as a substrate for sensitive and reproducible SERS based assays: application to the quantitation of malachite green and thiram. Microchim Acta. 2019;186(3):144. https://doi.org/10.1007/s00604-019-3257-4.
12. Deep A, Bhardwaj SK, Paul AK, Kim KH, Kumar P. Surface assembly of nano-metal organic framework on amine functionalized indium tin oxide substrate for impedimetric sensing of parathion. Biosens Bioelectron. 2015;65:226–31. https://doi.org/10.1016/J.BIOS.2014.10.045.
13. Wang B-H, Lian X, Yan B. Recyclable Eu3+ functionalized Hf-MOF fluorescent probe for urinary metabolites of some organophosphorus pesticides. Talanta. 2020;214:120856. https://doi.org/10.1016/j.talanta.2020.120856.
14. Luo L, et al. Rational construction of a robust metal-organic framework nanozyme with dual-metal active sites for colorimetric detection of organophosphorus pesticides. J Hazard Mater. 2022;423:127253. https://doi.org/10.1016/j.jhazmat.2021.127253.
15. Cai Y, et al. Capsulation of AuNCs with AIE effect into metal–organic framework for the marriage of a fluorescence and colorimetric biosensor to detect organophosphorus pesticides. Anal Chem. 2021;93(19):7275–82. https://doi.org/10.1021/acs.analchem.1c00616.
16. Wiwasuku T, et al. Turn-on fluorescent probe towards glyphosate and Cr $^{3+}$ based on Cd(ii)-metal organic framework with Lewis basic sites. Inorg Chem Front. 2021;8(4):977–88. https://doi.org/10.1039/D0QI00947D.
17. Xu X, et al. Sensitive detection of pesticides by a highly luminescent metal-organic framework. Sens Actuators B Chem. 2018;260:339–45. https://doi.org/10.1016/J.SNB.2018.01.075.
18. Li Y, Wu S, Zhang Y, Ma Z, Zhu M, Gao E. A lanthanide metal–organic framework as ratio fluorescence probe to detect pesticides in water. Inorganica Chim Acta. 2021;528:120632. https://doi.org/10.1016/J.ICA.2021.120632.
19. Wei W, Wang J, Tian C-B, Du S-W, Wu K-C. A highly hydrolytically stable lanthanide organic framework as a sensitive luminescent probe for DBP and chlorpyrifos detection. Analyst. 2018;143(22):5481–6. https://doi.org/10.1039/C8AN01606B.
20. Chen H, et al. A bifunctional luminescent metal–organic framework for the sensing of Paraquat and Fe $^{3+}$ Ions in water. Chem Asian J. 2019;14(20):3611–9. https://doi.org/10.1002/asia.201900682.
21. Fan L, et al. Two cadmium(II) coordination polymers as multi-functional luminescent sensors for the detection of Cr(VI) anions, dichloronitroaniline pesticide, and nitrofuran antibiotic in aqueous media. Spectrochim Acta A Mol Biomol Spectrosc. 2020;239:118467. https://doi.org/10.1016/J.SAA.2020.118467.
22. Zhao B, et al. An anionic-ligand installed pyrene-based MOF for the fluorescence detection of paraquat. New J Chem. 2021;45(9):4401–7. https://doi.org/10.1039/D0NJ05866A.
23. Wang X-Q, et al. A water-stable zinc(ii)–organic framework as a multiresponsive luminescent sensor for toxic heavy metal cations, oxyanions and organochlorine pesticides in aqueous solution. Dalton Trans. 2019;48(44):16776–85. https://doi.org/10.1039/C9DT03195B.

24. Bhardwaj SK, et al. Immunosensing of atrazine with antibody-functionalized cu-MOF conducting thin films. ACS Appl Mater Interfaces. 2015;7(47):26124–30. https://doi.org/10.1021/acsami.5b07692.
25. Feng D-D, et al. Two novel metal–organic frameworks based on pyridyl-imidazole-carboxyl multifunctional ligand: selective CO_2 capture and multiresponsive luminescence sensor. Dalton Trans. 2019;48(29):10892–900. https://doi.org/10.1039/C9DT01430F.
26. Li C, et al. A 3D hierarchical dual-metal–organic framework heterostructure up-regulating the pre-concentration effect for ultrasensitive fluorescence detection of tetracycline antibiotics. J Mater Chem C Mater. 2020;8(6):2054–64. https://doi.org/10.1039/C9TC05941E.

Metal-Organic Frameworks for Detection of Antibiotics in Forensic Samples

6

Abstract

In the realm of forensic science, the identification and quantification of antibiotics in various samples play a crucial role in investigations related to drug-related crimes, poisoning, and medication misuse. Metal-organic frameworks (MOFs) have garnered considerable attention in recent years for their exceptional properties and diverse applications, including sensing and detection. This chapter explores the potential of MOFs as cutting-edge platforms for the detection of antibiotics in forensic samples. It discusses the underlying principles of MOF-based sensors, their fabrication techniques, and the mechanisms governing antibiotic sensing. Moreover, specific examples of MOF-based sensors for the detection of antibiotics are elucidated, highlighting their sensitivity, selectivity, and applicability in forensic analysis. Furthermore, the chapter addresses challenges and future prospects in the development and implementation of MOF-based sensors in forensic investigations, underscoring their significance in enhancing analytical capabilities and advancing forensic science methodologies.

6.1 Introduction

Antibiotics are substances which retards the growth of microbes or kills them [1–3]. Though, antibiotics have immunosuppressive actions but inappropriate usage of antibiotics is the cause of anaphylatic death [4]. Body fluids like blood or urine and viscera are sent to Forensic Science Laboratory (FSL) to detect and quantify antibiotics [5–11]. There are also cases where animal feeds are sent to forensic labs to detect toxic levels of antibiotics [12].

Detection of antibiotics are carried out using analytical techniques like LC-MS, GC-MS, Capillary electrophoresis and immunoassay [13]. However, these

V. Mishra, A. Mishra, *Metal-Organic Frameworks as Forensic Detectors*,
https://doi.org/10.1007/978-981-96-2147-7_6

techniques are costly, time-consuming, and require sophisticated instrumentation [14]. Alternately, MOFs have emerged as promising techniques in detection of antibiotics with certain advantages like less time-consuming, high selectivity and senstivity (Fig. 6.1).

In 2016, [15] MOFs were initially used to detect antibiotics. Zr(IV)-MOF was used to detect 2,4,6- trinitrophenol with LOD of 90 ppb and Nitrofurazone with LOD of 58 ppb. Cd(II) based MOFs were used to detect Nitrofurazone. MOFs are also used to detect antibiotics like tetracycline, nitroimidazole and sulfonamide [16]. Zhou et al. detected tetracycline in water with LOD 30 nm using first water-stable PCN-128Y (Zr-based MOF) based on (H_4ETTC) and TPE ligand [17]. Li et al. explored (Al-MOF@Mo/Zn-MOF). The π-π interaction and hydrogen bonding between tetracycline and this MOF is due to strong chealting capability of metal node of Zn, Al, and Mo to analyte [18]. When In(III)- MOF is excited at 280 nm, it displays an emission at 380 nm. In(III)- MOF can detect fluoroquinoles (ciprofloxacin, norfloxacin and enfrofloxacin) with detection limits of 72.9, 56.7, and 79 as shown in Fig. 6.2.

The ability of various MOFs to detect antibiotics and their limit of detection are summarized in Table 6.1.

MOFs-based sensors exhibit excellent performance in the detection of antibiotics. However, because of difficulty in the pore modification, there is certain limitations with MOfs application [16].

This study aims to address growing concerns regarding the presence of antibiotic residues, specifically tetracyclines (TCs) such as tetracycline (TET),

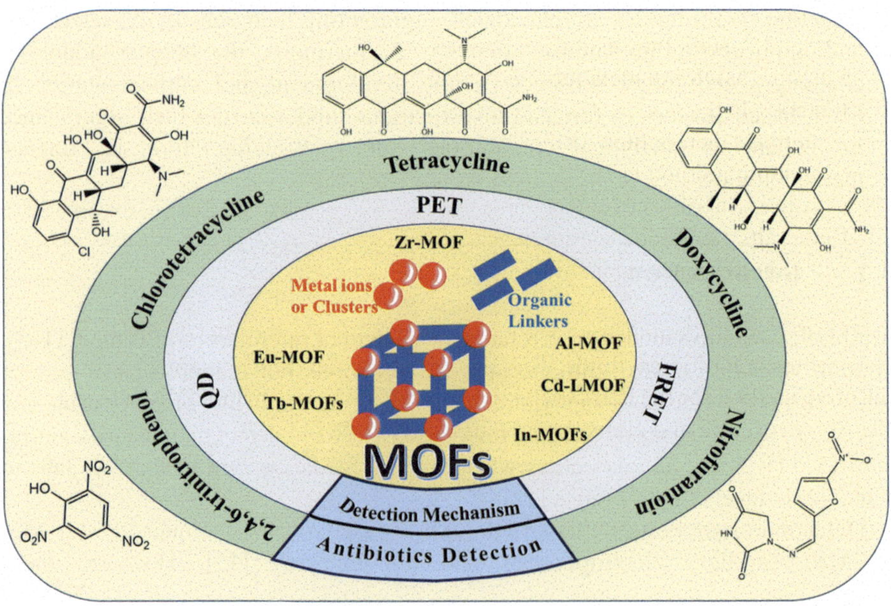

Fig. 6.1 MOFs for pesticide detection

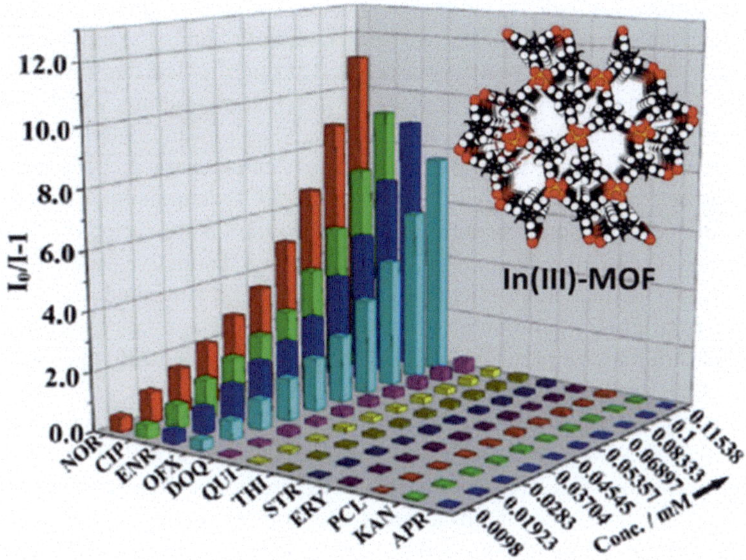

Fig. 6.2 Detection limit of fluoroquinoles. (Reprint with permission [19]. Copyright @2020 RSC)

Table 6.1 Ability of various MOFs to detect antibiotics with their detection limit

S. No.	MOFs	Antibiotics	Sensing category	Limit of detection	References
1.	Zr(IV)-MOF	NZF TNP	PET FRET	58 ppb	[15]
2.	In(III)-MOFs	Norfloxacin Enrofloxacin Ciprofloxacin	FRET PET PET	56.7 ppb 79 ppb 72.9 ppb	[19]
3.	Zr-based MOF (PCN-128Y)	Tetracycline	FRET PET	30 nM	[17]
4.	$\{[Cd_3(L)_2(bbi)_2(H_2O)_2]\cdot 2H_2O\}$	NZF	FRET	1.83 ppM	[20]
5.	$[Zn_3(\mu_3\text{-}OH)(HL)L(H_2O)_3]\cdot H_2O$	Sulfonamide	PET	–	[18]
6.	$\{[Zn_2(bcob)(OH)(H_2O)]\cdot DMA\}_n$ (ROD-Zn1)	Tetracycline	PET	0.11 μmol	[21]
7.	$[Cd_2Na(L)(BDC)_{2.5}]\cdot 9H_2O$	NZF	PET	162 ppb	[22]
8.	Cd-based LMOF	NFT	PET	0.26 μM	[23]
9.	Eu^{3+} and Tb^{3+} co-doped LMOFs	Minocycline Norfloxacin	FRET PET	1.23 μM 0.06 μM	[24]
10.	Al-MOF@Mo/Zn-MOF	DOX TET OTC CTC	FRET FRET PET PET	0.56 nM 0.53 nM 0.58 nM 0.86 nM	[25]

chlortetracycline (CTC), doxycycline (DOX), and oxytetracycline (OTC). It recognizes the detrimental impacts of these residues on both human health and the environment. The chapter presents a newly developed 3D hierarchical dual-metal–organic framework, known as the Al-MOF@Mo/Zn-MOF heterostructure, which is synthesized via an internal extended growth technique. The present study highlights the exceptional capabilities of this novel material in detecting trace concentrations of TCs with great sensitivity. The Al-MOF@Mo/Zn-MOF demonstrates excellent fluorescence quenching upon exposure to TCs due to the identification of TCs by the organic ligands included in the material, which possess distinctive functional groups such as $-NH_2$. The quenching phenomenon seen may be attributed to a synergistic interplay between hydrogen bonding interactions, which facilitate electron transfer between the amino groups ($-NH_2$) on the ligands, and the carbonyl (-CO-) and hydroxyl (-OH) groups on the transition metals (TCs).

Additionally, the significant absorption of the transition metals at the excitation energy of the Al-MOF@Mo/Zn-MOF contributes to this effect. The exceptional sensitivity of the material can be attributed to its pre-concentration effect, which is facilitated by the presence of unsaturated sites on the nodes of the dual-metal-organic framework. These unsaturated sites allow for the recognition and adsorption of nitrogen and oxygen atoms from the TC molecules. Additionally, the hierarchical nanoarchitecture of the material promotes mass transfer and reduces diffusion barriers, resulting in an increased contact area between the MOF and TC molecules. Consequently, the Al-MOF@Mo/Zn-MOF has enhanced sensitivity, with very low limits of detection (LODs) of 0.56, 0.53, 0.58, and 0.86 nM for DOX, TET, OTC, and CTC, respectively. The limits of detection (LODs) shown by these LODs are much lower by two orders of magnitude compared to the LODs of pure amino-functionalized Al-MOF. Furthermore, they are like, and in some cases even exceed, the LODs of earlier sensors that rely on fluorescence as the detection mechanism. The Al-MOF@Mo/Zn-MOF exhibits remarkable resistance to interference and produces adequate recovery rates ranging from 87.07% to 116.44%. Additionally, the method displays low relative standard deviations (RSDs) between 0.22% and 2.30% when used for the analysis of actual water and food samples. The 3D hierarchical Al-MOF@Mo/Zn-MOF heterostructure exhibits remarkable sensing capabilities, along with its convenient manufacturing process, stability, and practical feasibility. Consequently, it emerges as a highly promising contender for ultrasensitive temperature coefficient (TC) detection in many real-world applications [25].

6.2 Detection of 2,4,6-Trinitrophenol

A solvothermal approach was used to create the amine-decorated zirconium-based metal organic framework (MOF) UiO-66-NH_2, which has a rod-shaped morphology (Fig. 6.3). 2-Aminoterephthalic acid was used as an organic linker in the procedure. Using PXRD, the produced MOF material's crystallinity was verified. For the ultra-trace detection of 2,4,6-trinitrophenol (TNP) in an aqueous matrix, even in the

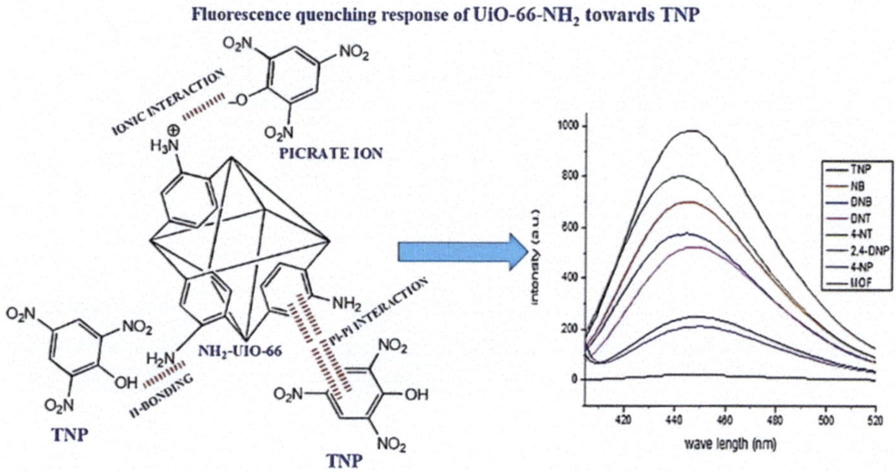

Fig. 6.3 Graphical response of UiO-66-NH2 towards TNP. (Reprint with permission [26], Copyright @2023 Springer)

presence of other competitive nitroaromatic analytes, MOF was utilized as a sensitive and selective sensor.

The developed MOF for TNP was shown to have outstanding sensitivity based on its high value of the Stern-Volmer quenching constant Ksv ($1.106 \times 10^5\ M^{-1}$), credible photoluminescent quenching efficiency (97.8%), and lower detection limit (0.95 μM/217 ng mL^{-1}). Electrostatic interactions (i.e., ionic interaction, H-bonding, and π-π interaction) and density functional theory calculations revealed that electron and energy transfer mechanisms are important in the UiO-66-NH_2 sensor's turn-off quenching response [26].

The porous materials known as metal-organic frameworks (MOFs) and covalent-organic frameworks (COFs) have shown great promise in applications such as sensing, catalysis, drug delivery, water generation, gas capture, and pollution removal. The adjustable characteristics of MOFs and COFs, like their appealing morphologies, active functional sites, and ultra-trace porosity, make them an excellent substrate for ensnaring a variety of analytes from clinical, environmental, and food samples. Furthermore, MOFs and COFs have demonstrated utility in the study of pharmacological residues [27] (Fig. 6.4).

6.3 Detection of Tetracycline

The presence of tetracycline antibiotics and dihydrogen phosphate ($H_2PO_4^-$) in food and the environment poses significant risks to human health. This study introduces a water-stable europium-based metal–organic framework (Eu-MOF), which is functionalized with carboxyl groups, as a novel dual-responsive fluorescent sensor for detecting tetracycline antibiotics (specifically TET, OTC, and DOX) and $H_2PO_4^-$.

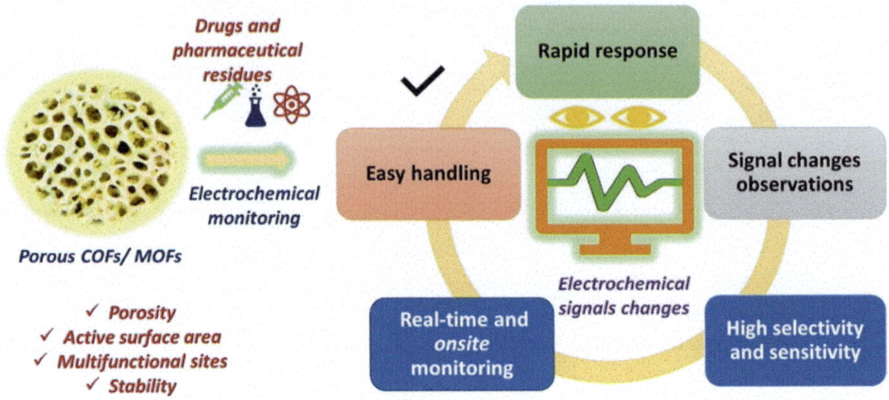

Fig. 6.4 Representation of beneficial facts by MOFs in detection of antibiotics. (Reprint with permission [27], Copyright @2023 Elsevier)

This sensor exhibits fluorescence turn-on for tetracyclines and fluorescence turn-off for dihydrogen phosphate. Eu-MOF provides sensitive and selective detection of oxytetracycline (OTC) with a fast response time of just 1 m and high resistance to interference. The limits of detection (LODs) were calculated at 78 nM for OTC, 225 nM for TET, and 201 nM for DOX. The fluorescence enhancement mechanism involves energy and electron transfer from the tetracycline molecules to the MOF, facilitated by coordination and hydrogen bonding. Additionally, the Eu-MOF effectively detects $H_2PO_4^-$ by quenching fluorescence, with an LOD of 0.70 μM. This quenching occurs due to interference with the energy transfer between the ligand and the MOF. The sensor was successfully applied to detect OTC and $H_2PO_4^-$ in real samples, with excellent recoveries and low relative standard deviations (RSDs).

Moreover, this Eu-MOF has a unique application in developing latent fingerprints on various surfaces, offering clear fluorescent fingerprint details, with sweat pores visible to the naked eye. [28] (Fig. 6.5).

6.4 Determination of Lincomycin Using Cu-Based Metal-Organic Framework

Residues from antibiotic drugs may have negative effects on human health. Common veterinary medication lincomycin might leave residues in foods that come from animals. It can be difficult to find traces of lincomycin residues in actual samples, though. Here, lincomycin was extracted using a basic solid phase extraction (SPE) technique and then enriched in cow milk samples in order to be detected by high-performance liquid chromatography-tandem mass spectrometry (HPLC-MS/MSA). A Cu-based metal-organic framework (Cu-MOF) synthesized through the solvothermal method was used as an adsorbent in solid-phase extraction (SPE). To characterize the resulting MOFs, several analytical techniques were employed, including scanning electron microscopy (SEM) (Fig. 6.6), Fourier-transform infrared

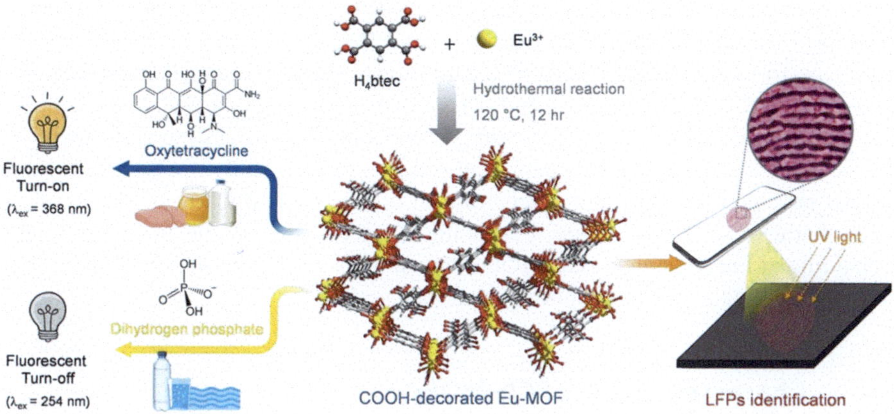

Fig. 6.5 Schematic representation for the synthesis of water-stable Eu-MOF and the detection of oxytetracycline and H_2PO_4. (Reprint with permission [28], Copyright @2023 RSC)

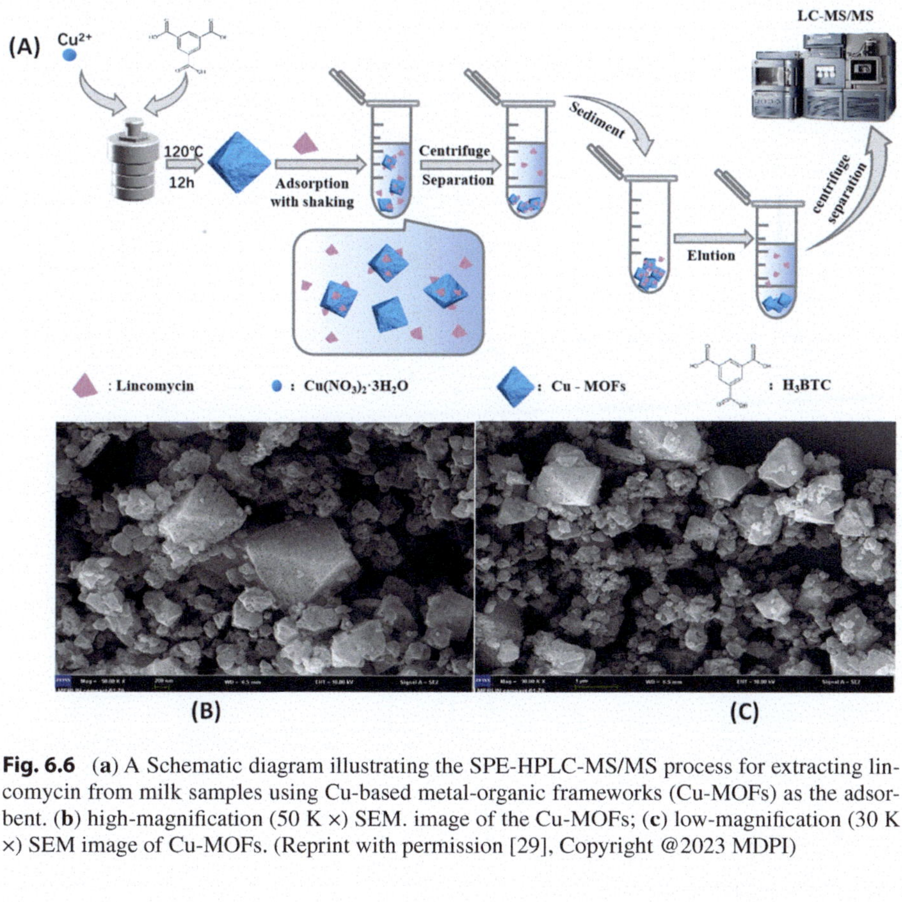

Fig. 6.6 (**a**) A Schematic diagram illustrating the SPE-HPLC-MS/MS process for extracting lincomycin from milk samples using Cu-based metal-organic frameworks (Cu-MOFs) as the adsorbent. (**b**) high-magnification (50 K ×) SEM. image of the Cu-MOFs; (**c**) low-magnification (30 K ×) SEM image of Cu-MOFs. (Reprint with permission [29], Copyright @2023 MDPI)

spectroscopy (FT-IR), X-ray diffractometry (XRD), differential thermogravimetric analysis (TG-DTA), and nitrogen adsorption-desorption tests. Various factors affecting the performance of the Cu-MOF, such as the elution solvent, adsorption capacity (equilibrium), extraction time, and pH, were also investigated [29]. Coordination polymers allow Cu-based metal-organic frameworks (Cu-MOFs) to self-assemble from copper ions, oxygen, nitrogen, and other organic ligands. Cu-MOFs exhibit distinct pore architectures and a spectrum of adsorption, optical, and electromagnetic characteristics due to the well-defined orientations of the metal ions and organic ligand arrangement. Cu-MOFs thus show a lot of promise and appealing futures in the realm of contemporary materials research.

6.5 Characterization of Cu-MOFs

Cu-MOFs' morphology was assessed using scanning electron microscopy (SEM). The Cu-MOF has a surface pore structure and a conventional octahedral form. The figure below displays the Cu-MOFs layer's XRD pattern as it was constructed (Fig. 6.7). The following XRD pattern (Fig. 6.7) illustrates the formation of the

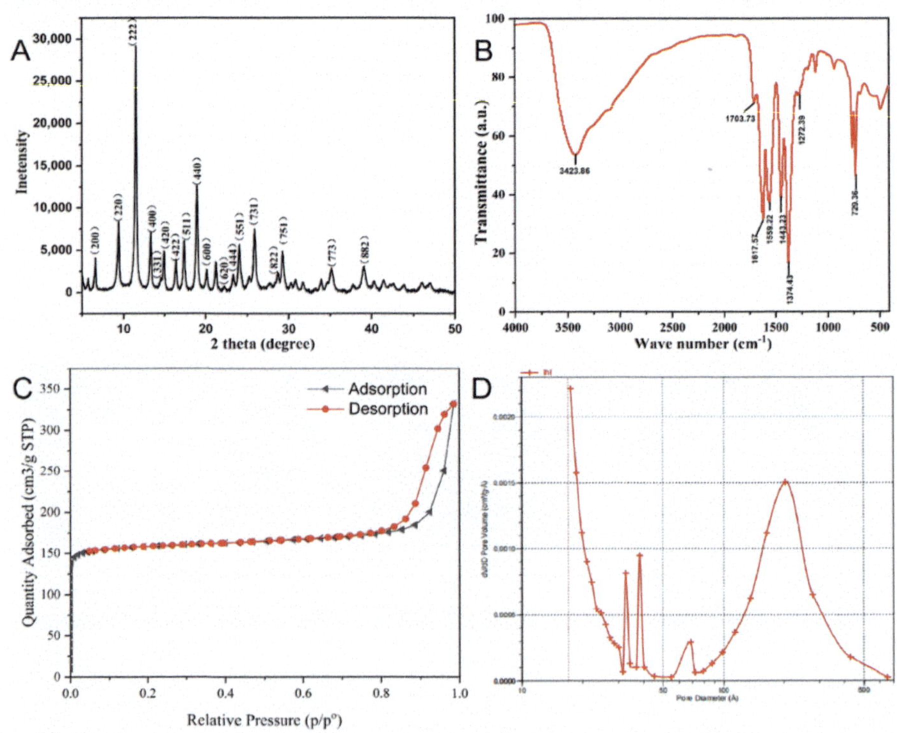

Fig. 6.7 (**a**) XRD patterns of Cu–MOFs, (**b**) FT-IR spectra of Cu–MOFs, (**c**) N_2 adsorption-desorption isotherms of Cu–MOFs, (**d**) Pore size distribution of Cu–MOFs. (Reprint with permission [29], Copyright @2023 MDPI)

Cu-MOFs layer. The Cu-MOFs synthesized here exhibit a face-centered cubic (FCC) crystal lattice. Distinct reflections are observed at 2θ values of 11.56°, 13.38°, 14.54°, 14.96°, 16.38°, 17.42°, 18.96°, 20.14°, 23.32°, 24.06°, 25.86°, 28.62°, 29.26°, 35.16°, and 39.1°.

These correspond to the (222), (400), (331), (420), (422), (511), (440), (600), (444), (551), (731), (822), (751), (773), and (882) crystallographic planes, aligning with previously published studies. These findings suggest the successful synthesis of Cu-MOFs crystals with the expected topological structure [29].

Cu-MOFs were made using the solvothermal technique, and they were successfully used for the first time to find lincomycin in milk samples. Lincomycin had an adsorption capacity of 131.41 mg/g, which was in accordance with the pseudo-second-order adsorption kinetics model and the Langmuir isothermal adsorption equation. Cu-MOFs demonstrated excellent extraction efficiency, enrichment, and stability at pH = 7, with the best adsorption performance at that pH. Good precision, a low detection limit, a high recovery rate, and ease of use were all benefits of the procedure. Cu-MOF is therefore a potential adsorbent material that can be used to extract and enhance other antibiotics from various materials [29].

6.6 Luminescence and Turn on Sensing in MOF

When exposed to a specific wavelength of light, two types of light can be produced: fluorescence and phosphorescence. This process is known as luminescence. The molecule undergoes a radiative transition, or jumps from the ground state to the excited state upon absorbing incident light, and subsequently returns to the ground state either by fluorescence or phosphorescence. The fluorescence happens between energy levels with equal spin multiplicity ($S^1 \rightarrow S^0$) and happens very quickly (10–15 s). On the other hand, phosphorescence is a type of delayed luminescence that typically takes a few microseconds to a few seconds. The emission wavelength of phosphorescence occurs between energy levels with distinct spin multiplicities ($T^1 \rightarrow S^0$) [30].

It is evident that MOF has the ability to either "turn on" or "turn off" fluorescence in order to detect the target analyte. The MOF experiences a luminescence boost while using the "turn-on" procedure, while the MOF experiences a decrease in fluorescence intensity when experiencing a "turn-off" event. Surprisingly, the former is better for applications involving real-time sensing. Unlike the quenching (turn-off) action, luminescence amplification occurs against a dark background and is therefore simpler to observe. Furthermore, interference analytes rather than the target analyte may be the origin of the "turn-off" event. This is the primary barrier to real-time application and frequently results in incorrect predictions [31].

Furthermore, as a luminescence [34] boost is needed in the dark backdrop, the "turn-on" procedure is highly desired in biological systems. As a result, the analyte detection can be easily monitored thanks to emission amplification. In contrast, the "turn-off" event lacks this encouraging information. The application of MOFs [32] as fluorescence "turn-on" probes for diverse analyte detection is described in the sections that follow [31, 33].

References

1. Anand R, Yadav N, Mudgal D, Jindal S, Sengupta S, Kumar D, et al. Synthesis, in-silico molecular docking studies, and in-vitro antimicrobial evaluation of Isatin scaffolds bearing 1, 2, 3-triazoles using click chemistry. Indian J Microbiol. 2024; https://doi.org/10.1007/s12088-024-01264-z.
2. Etebu E, Arikekpar I. Antibiotics: classification and mechanisms of action with emphasis on molecular perspectives. Int J Appl Microbiol Biotechnol Res. 2016;4(2016):90–101.
3. Yadav N, Mudgal D, Mishra and V. Nanobiotic formulations utilizing quinoline-based-triazole functionalized carbon quantum dots via click chemistry for combatting clinical-resistant bacterial pathogens. Indian J Microbiol. 2024; https://doi.org/10.1007/s12088-024-01266-x.
4. Shen Y, Li L, Grant J, Rubio A, Zhao Z, Zhang X, et al. Anaphylactic deaths in Maryland (United States) and Shanghai (China): a review of forensic autopsy cases from 2004 to 2006. Forensic Sci Int. 2009;186(1–3):1–5.
5. Behl A, Mishra A, Sharma GP, Mishra V. Detection and determination of the levels of physiologically active substances in non-alcoholic beverages. Int J Med Toxicol Leg Med. 2021;25(3–4):164–74.
6. Chaudhary J, Mavi AK, Singh N, Srivastava VK, Jyoti A, Kaushik S, et al. Exploring the binding of Ajmalicine, dexamethasone and acetaminophen to enterococcus faecalis Homoserine dehydrogenase. Quad J Quaderns J. 2023;11(3):294–307. (https://quadernsorg/volume-11-issue-3-2023/
7. Mishra A, Mishra V, Srivastava A. Improved isolation and detection of Malathion insecticide from forensic samples. Int J Med Toxicol Leg Med. 2021a;25(3–4):148–54.
8. Mishra A, Sai M, Rani S, Singh C, Mishra V. Extraction, isolation, and detection methods used for antihistamines drugs from biological matrices- a review. Int J Med Toxicol Leg Med. 2021b;25(3–4):155–63.
9. Mishra A, Srivastava A, Kanojia R, Mishra V. Improved isolation and detection approach of chlorpyrifos pesticide from urine and viscera samples. J Forensic Med Toxicol. 2022;39(1):45–50.
10. Ngo HL, Mishra DK, Mishra V, Truong CC. Recent advances in the synthesis of heterocycles and pharmaceuticals from the photo/electrochemical fixation of carbon dioxide. Chem Eng Sci. 2021;229:116142.
11. Srivastava A, Mishra V, Singh P, Kumar R. Coumarin-based polymer and its silver nanocomposite as advanced antibacterial agents: synthetic path, kinetics of polymerization, and applications. J Appl Polym Sci. 2012;126(2):395–407.
12. Ebel JG, Wachs T, Henion JD. Rapid forensic selected reaction monitoring liquid chromatography/mass spectrometry determination of ionophore antibiotics found at toxic levels in animal feeds. J AOAC Int. 2004;87(1):25–30.
13. Dabodhia KL, Tripathi B, Lamba NP, Chauhan MS, Bhatia R, Mishra and V. Optimization and bioassay guided comparative techniques for efficient extraction of lutein esters from Tagetes erecta (Var. Pusa Narangi Genda) flowers. Mass Spectrom Lett Korean Soc Mass Spectrom. 2024;15(1):40–8.
14. Zhao D, Yu S, Jiang WJ, Cai ZH, Li DL, Liu YL, et al. Recent progress in metal-organic framework based fluorescent sensors for hazardous materials detection. Molecules. 2022:2226.
15. Wang B, Lv X-L, Feng D, Xie L-H, Zhang J, Li M, et al. Highly stable Zr (IV)-based metal–organic frameworks for the detection and removal of antibiotics and organic explosives in water. J Am Chem Soc. 2016;138(19):6204–16.
16. Tao C-L, Chen B, Liu X-G, Zhou L-J, Zhu X-L, Cao J, et al. A highly luminescent entangled metal–organic framework based on pyridine-substituted tetraphenylethene for efficient pesticide detection. Chem Commun. 2017;53(72):9975–8.
17. Zhou Y, Yang Q, Zhang D, Gan N, Li Q, Cuan J. Detection and removal of antibiotic tetracycline in water with a highly stable luminescent MOF. Sensors Actuators B Chem. 2018;262:137–43.

18. Zhu X-D, Zhang K, Wang Y, Long W-W, Sa R-J, Liu T-F, et al. Fluorescent metal–organic framework (MOF) as a highly sensitive and quickly responsive chemical sensor for the detection of antibiotics in simulated wastewater. Inorg Chem. 2018;57(3):1060–5.
19. Zhong W-B, Li R-X, Lv J, He T, Xu M-M, Wang B, et al. Two isomeric in (III)-MOFs: unexpected stability difference and selective fluorescence detection of fluoroquinolone antibiotics in water. Inorg Chem Front R Soc Chem. 2020;7(5):1161–71.
20. Zhou S, Lu L, Liu D, Wang J, Sakiyama H, Muddassir M, et al. Series of highly stable Cd (ii)-based MOFs as sensitive and selective sensors for detection of nitrofuran antibiotic. Cryst Eng Comm R Soc Chem. 2021;23(46):8043–52.
21. Gai S, Zhang J, Fan R, Xing K, Chen W, Zhu K, et al. Highly stable zinc-based metal–organic frameworks and corresponding flexible composites for removal and detection of antibiotics in water. ACS Appl Mater Interf. 2020;12(7):8650–62.
22. Zhao D, Liu X-H, Zhao Y, Wang P, Liu Y, Azam M, et al. Luminescent Cd (II)–organic frameworks with chelating NH 2 sites for selective detection of Fe (III) and antibiotics. J Mater Chem A R Soc Chem. 2017;5(30):15797–807.
23. Li P, Guo M-Y, Gao L-L, Yin X-M, Yang S-L, Bu R, et al. Photoresponsivity and antibiotic sensing properties of an entangled tris (pyridinium)-based metal–organic framework. Dalt Trans. 2020b;49(22):7488–95.
24. Xie R, Yang P, Liu J, Zou X, Tan Y, Wang X, et al. Lanthanide-functionalized metal-organic frameworks based ratiometric fluorescent sensor array for identification and determination of antibiotics. Talanta. 2021;231:122366.
25. Li C, Yang W, Zhang X, Han Y, Tang W, Yue T, et al. A 3D hierarchical dual-metal–organic framework heterostructure up-regulating the pre-concentration effect for ultrasensitive fluorescence detection of tetracycline antibiotics. J Mater Chem C R Soc Chem. 2020a;8(6):2054–64.
26. Verma R, Dhingra G, Kaur M, Garg D, Mohiuddin I, Malik AK. Amine-decorated zirconium based metal organic framework for ultrafast detection of 2,4,6-trinitrophenol in aqueous samples. J Fluoresc. 2023 Sep;33(5):2085–98.
27. Mohan B, Kumari R, Virender SG, Singh K, Pombeiro AJL, et al. Covalent organic frameworks (COFs) and metal–organic frameworks (MOFs) as electrochemical sensors for the efficient detection of pharmaceutical residues. Environ Int. 2023:107928.
28. Wiwasuku T, Chuaephon A, Puangmali T, Boonmak J, Ittisanronnachai S, Promarak V, et al. Multifunctional fluorescent Eu-MOF probe for tetracycline antibiotics and dihydrogen phosphate sensing and visualizing latent fingerprints. RSC Adv. 2023;13(15):10384–96.
29. Li H, Wu J, Bai J, Wu J, Wu J. Determination of lincomycin in Milk using cu-based metal-organic framework adsorbent and liquid chromatography-tandem mass spectrometry. Molecules. 2023;28(14):5307.
30. Feng JY, Yang YH, Su YQ, Dong GY. A zinc(II)-based 3D metal−organic framework with rare dmc topology as a dual ratiometric and turn-off luminescent sensor. Inorg Chem Commun. 2023;156:111247.
31. Pal TK. Metal-organic framework (MOF)-based fluorescence "turn-on" sensors. Mater Chem Front R Soc Chem. 2022:405–41.
32. Dwivedi B, Mishra PK, Mishra V. Chiral metal-organic frameworks for asymmetrical catalysis. In Metal-organic frameworks (MOFs) as catalysts, Singapore: Springer Nature Singapore, 2022;493–513.
33. Yadav N, Mudgal D, Mishra A, Shukla S, Malik T, Mishra V. Harnessing fluorescent carbon quantum dots from natural resource for advancing sweat latent fingerprint recognition with machine learning algorithms for enhanced human identification. Plos One. 2024;19(1):e0296270.
34. Guliani E, Taneja A, Ranjan KR, Mishra V. Luminous insights: Exploring organic fluorescent "Turn-On" chemosensors for metal-ion (Cu+ 2, Al+ 3, Zn+ 2, Fe+ 3) detection. J Fluoresc. 2024;34(5):1965–2001.

Metal-Organic Frameworks for Detection of Drugs of Abuse

7

Abstract

Metal-organic frameworks (MOFs) have emerged as a promising class of materials for the detection of drugs of abuse due to their high selectivity and adsorption capacity. The unique properties of MOFs, such as their high surface area and tunable pore sizes, enable the selective adsorption of target molecules, making them ideal for the detection of drugs of abuse. MOFs can be functionalized with different ligands to enhance their selectivity towards specific molecules, enabling the development of highly sensitive and selective methods for drug detection. In recent years, several studies have demonstrated the potential of MOFs for the detection of drugs of abuse. For instance, ZIF-8 has been functionalized with a specific antibody for the detection of human chorionic gonadotropin (hCG) in urine samples, while NH2-MIL-125(Ti) has been employed as a stationary phase in High-Performance Liquid Chromatography (HPLC) for the separation of antipsychotics in human serum. These studies demonstrate the potential of MOFs for the detection of drugs of abuse. The integration of MOFs with other analytical techniques, such as mass spectrometry or chromatography, also offers a promising approach for drug detection. MOFs can be used as extraction phases in sample preparation, enabling the selective extraction of target molecules from complex matrices. The combination of MOFs with other analytical techniques enables the development of highly sensitive and selective methods for drug detection and analysis. Despite the challenges in the application of MOFs for drug detection and analysis, such as their stability and reproducibility, the potential of MOFs in this field is still being explored, and further research in this area is necessary to fully realize their potential. Overall, MOFs offer a promising approach for the detection of drugs of abuse, and the integration of MOFs with other analytical techniques provides a pathway for the development of highly sensitive and selective methods for drug detection and analysis.

V. Mishra, A. Mishra, *Metal-Organic Frameworks as Forensic Detectors*,
https://doi.org/10.1007/978-981-96-2147-7_7

7.1 Introduction

7.1.1 Drugs of Abuse

Drugs of abuse are substances that can be used to alter one's mood or behaviour. They can be legal or illegal, and they can be used for a variety of purposes, including recreation, pain relief, and sedation. However, when used improperly, drugs of abuse can have dangerous and even deadly consequences.

7.1.2 Examples of Drugs of Abuse

There are many different types of drugs of abuse, but some of the most common include:

- Alcohol: Alcohol is a depressant that slows down the central nervous system. It can cause intoxication, slurred speech, impaired coordination, and memory loss. Long-term use of alcohol can lead to addiction, liver damage, and heart disease (Fig. 7.1).
- Nicotine: Nicotine is a stimulant that speeds up the central nervous system. It is the addictive substance in tobacco products, such as cigarettes, cigars, and chewing tobacco. Nicotine can cause addiction, lung cancer, and heart disease (Fig. 7.2).
- Marijuana: Marijuana is a psychoactive drug that comes from the Cannabis sativa plant. It can cause euphoria, relaxation, and altered perception. Long-term use of marijuana can lead to addiction, respiratory problems, and impaired cognitive function (Fig. 7.3).

Fig. 7.1 Alcohol [1]

Fig. 7.3 Marijuana [3]

Fig. 7.2 Nicotine [2]

- Cocaine: Cocaine is a stimulant that is derived from the coca plant. It can cause euphoria, increased energy, and alertness. However, cocaine is also a highly addictive and dangerous drug. It can cause heart attack, stroke, and seizures (Fig. 7.4).
- Heroin: Heroin is an opioid that is derived from the opium poppy. It is a powerful depressant that can cause intense euphoria, relaxation, and pain relief. However, heroin is also highly addictive and dangerous. It can lead to overdose, respiratory depression, and death (Fig. 7.5).
- Methamphetamine: Methamphetamine is a stimulant that is chemically similar to amphetamine. It can cause euphoria, increased energy, and alertness. However, methamphetamine is also highly addictive and dangerous. It can cause psychosis, heart problems, and stroke (Fig. 7.6).

7.1.3 Properties of Drugs of Abuse

Drugs of abuse can have a variety of properties, including:

- Addictiveness: The potential for a drug to cause dependence and addiction.
- Tolerance: The need to take increasing amounts of a drug to achieve the same effect.

Fig. 7.4 Cocaine [4]

Fig. 7.5 Heroin [5]

- Withdrawal: The physical and psychological symptoms that occur when a person stops using a drug.
- Toxicity: The potential for a drug to cause harm or death.

7.1.4 Effects of Drugs of Abuse

Drugs of abuse can have a variety of effects on the body and mind, including:

- Intoxication: The impairment of one's mental and physical faculties.
- Addiction: A chronic disease characterized by compulsive drug use despite harmful consequences.
- Overdose: The ingestion of a harmful amount of a drug, which can lead to death.

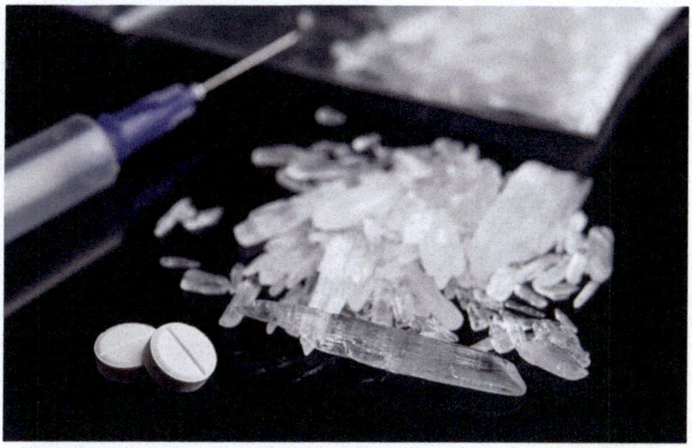

Fig. 7.6 Methamphetamine [6]

- Poisoning: The ingestion of a harmful substance, which can lead to illness or death.
- Death: Drugs of abuse can cause death in a number of ways, including overdose, poisoning, and injury.

7.1.5 Prevention of Drug Abuse

There are several things that can be done to prevent drug abuse, including:

- Education: Teaching children and adolescents about the dangers of drug abuse.
- Treatment: Providing treatment for people who are struggling with drug abuse.
- Law enforcement: Enforcing laws against the manufacture, distribution, and possession of illegal drugs.

7.1.6 Detection Methods

There are various methods available for the detection of drugs of abuse, including immunoassays, chromatography, spectrometry, and biosensors.

Immunoassays are one of the most commonly used methods for the detection of drugs of abuse. These assays use antibodies to detect the presence of a specific drug or drug metabolite in a biological sample such as urine, blood, or saliva. Immunoassays provide rapid and sensitive detection of drugs of abuse and are commonly used in drug testing programs.

Chromatography is another method used for the detection of drugs of abuse. It separates the components of a mixture based on their physical and chemical properties and provides a qualitative and quantitative analysis of the target analyte. Gas Chromatography (GC) and Liquid Chromatography (LC) are the two most commonly used chromatography techniques for the detection of drugs of abuse.

Spectrometry is a technique that measures the interaction of electromagnetic radiation with the target analyte. It provides information about the chemical composition and structure of the target analyte and is widely used for the detection of drugs of abuse.

Techniques such as Infrared (IR) spectroscopy, Ultraviolet-Visible (UV-Vis) spectroscopy, and Mass Spectrometry (MS) are commonly used for the detection of drugs of abuse. Biosensors are another type of sensor that detects the presence of drugs of abuse by using biological materials such as enzymes, antibodies, or cells. These sensors provide high sensitivity and selectivity towards the target analyte and are widely used for the detection of drugs of abuse in forensic and clinical settings.

In conclusion, the detection of drugs of abuse is important for prevention, diagnosis, and treatment of addiction, as well as for law enforcement purposes. There are various methods available for the detection of drugs of abuse, including immunoassays, chromatography, spectrometry, and biosensors. Each method has its advantages and disadvantages and should be selected based on the specific application and target analyte.

Drug abuse is a major public health concern worldwide, and the development of sensitive and selective methods for the detection of drugs of abuse is essential for effective drug control and prevention. Metal-Organic Frameworks (MOFs) have emerged as a promising class of materials for drug detection due to their unique properties, including high surface area and tunable pore sizes. These properties enable the selective adsorption of target molecules, making MOFs ideal for the detection of drugs of abuse. MOFs can be functionalized with different ligands to enhance their selectivity towards specific molecules, enabling the development of highly sensitive and selective methods for drug detection. The integration of MOFs with other analytical techniques, such as mass spectrometry or chromatography, also offers a promising approach for drug detection and analysis. The use of MOFs for drug detection and analysis has the potential to revolutionize drug control and prevention efforts and offers a promising approach for the development of highly sensitive and selective methods for drug detection.

7.1.7 Fabrication of MOF-Based Sensors for Drug Detection

Metal-Organic Frameworks (MOFs) are 3D crystalline materials that are composed of metal ions and organic molecules. Due to their high surface area and tunable properties, MOFs have been extensively studied for various applications including gas storage, separation, catalysis, and sensing. Among these applications, MOF-based sensors for drug detection have gained significant attention due to their high sensitivity, selectivity, and low limit of detection. The fabrication of MOF-based sensors for drug detection typically involves the immobilization of the MOF on a suitable substrate followed by the incorporation of a transducer that can detect the target analyte.

The sensing mechanism is usually based on the interaction between the target molecules and the MOF, which results in a change in the electrical, optical, or

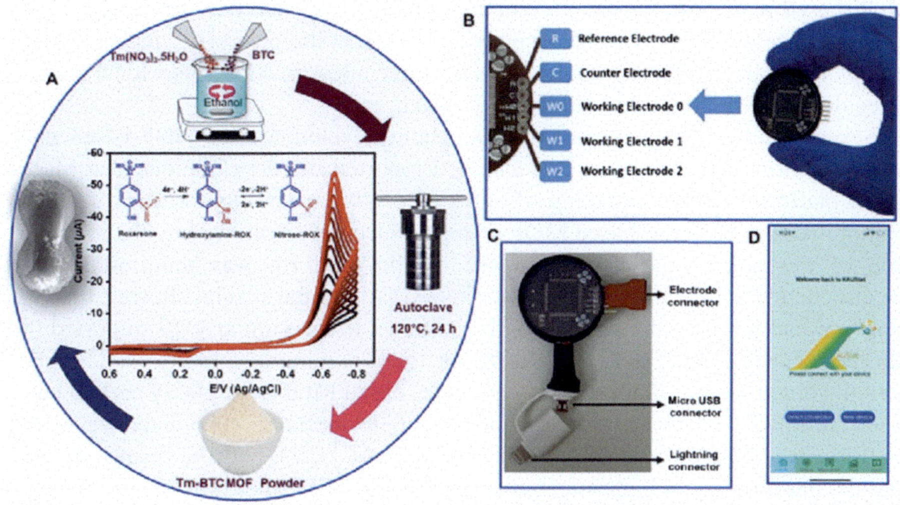

Fig. 7.7 Sensors for drug detection

mechanical properties of the sensor. Several studies have reported the successful fabrication of MOF-based sensors for the detection of various drugs including cocaine, heroin, and methamphetamine [7]. For instance, reported the fabrication of a MOF-based sensor for the detection of cocaine using a copper-based MOF and a Quartz Crystal Microbalance (QCM) transducer. The sensor showed a high selectivity towards cocaine with a limit of detection of 1.5 ng/mL. Similarly, [8] reported the fabrication of a MOF-based sensor for the detection of methamphetamine using a zirconium-based MOF and a Field-Effect Transistor (FET) transducer. The sensor showed a high sensitivity towards methamphetamine with a limit of detection of 1.8 ng/mL.

One of the key advantages of MOF-based sensors is their high selectivity towards specific target molecules. MOFs can be designed to have specific pore sizes and functional groups that can selectively bind to certain molecules, making them ideal for the detection of drugs and other hazardous substances. In conclusion, MOF-based sensors are a promising technology for drug detection due to their high sensitivity, selectivity, and low limit of detection. The fabrication of MOF-based sensors involves the immobilization of the MOF on a suitable substrate and the incorporation of a transducer. Several studies have reported the successful fabrication of MOF-based sensors for the detection of various drugs, highlighting the potential of this technology for real-world applications (Fig. 7.7).

7.1.8 Development of MOF-Based Platforms for Drug Screening

MOF-based platforms for drug screening typically involve the immobilization of the MOF on a substrate followed by the introduction of the target analyte, which can then be detected using various analytical methods. One of the key advantages of

MOF-based platforms for drug screening is their high selectivity towards specific target molecules. MOFs can be designed to have specific pore sizes and functional groups that can selectively bind to certain molecules, making them ideal for the screening of drugs and other hazardous substances.

Several studies have reported the successful development of MOF-based platforms for drug screening. For instance, [9] reported the development of a MOF-based platform for the screening of anti-cancer drugs. The platform involved the immobilization of a zinc-based MOF on a gold electrode, followed by the introduction of the target drug. The response of the platform was monitored using Electrochemical Impedance Spectroscopy (EIS), and the results showed that the platform had a high sensitivity towards the target drug. Similarly, [9] reported the development of a MOF-based platform for the screening of antibiotics. The platform involved the immobilization of a copper-based MOF on a glassy carbon electrode, followed by the introduction of the target antibiotic. The response of the platform was monitored using Cyclic Voltammetry (CV), and the results showed that the platform had a high selectivity towards the target antibiotic.

The development of MOF-based platforms typically involves the immobilization of the MOF on a substrate followed by the introduction of the target analyte, which can then be detected using various analytical methods. Several studies have reported the successful development of MOF-based platforms for the screening of various drugs, highlighting the potential of this technology for real-world applications.

7.1.9 MOFs for the Detection of Opioids, Cannabinoids, and Other Drugs of Abuse

Metal-Organic Frameworks (MOFs) have emerged as promising materials for the detection of opioids, cannabinoids, and other drugs of abuse due to their high sensitivity, selectivity, and low limit of detection. MOFs are composed of metal ions and organic ligands, and their tunable properties make them ideal for the detection of various target molecules. Several studies have reported the successful detection of opioids using MOFs. For instance, [10] reported the detection of fentanyl using a zirconium-based MOF and a Surface-Enhanced Raman Scattering (SERS) technique. The sensor showed a high sensitivity towards fentanyl with a limit of detection of 0.1 ng/mL. Similarly, [11] reported the detection of morphine using a copper-based MOF and a colorimetric method. The sensor showed a high selectivity towards morphine with a limit of detection of 0.94 ng/mL. MOFs have also been used for the detection of cannabinoids. For example, [12] reported the detection of Tetrahydrocannabinol (THC) using a zirconium-based MOF and a fluorescence quenching method. The sensor showed a high sensitivity towards THC with a limit of detection of 1.4 ng/mL. In addition to opioids and cannabinoids, MOFs have also been used for the detection of other drugs of abuse. For instance, [13] reported the detection of methamphetamine using a zirconium-based MOF and a fluorescence quenching method. The sensor showed a high sensitivity towards methamphetamine with a limit of detection of 0.8 ng/mL.

In conclusion, MOFs are promising materials for the detection of opioids, cannabinoids, and other drugs of abuse due to their high sensitivity, selectivity, and low limit of detection. Several studies have reported the successful detection of various drugs using MOFs, highlighting their potential for real-world applications.

7.1.10 MOFs for the Detection of Drug Metabolites in Biological Fluids

Metal-organic frameworks (MOFs) have great potential for the detection of drug metabolites in biological fluids due to their high sensitivity, selectivity, and low limit of detection. MOFs are composed of metal ions and organic ligands, and their tunable properties make them ideal for the detection of various target molecules. Several studies have reported the successful detection of drug metabolites in biological fluids using MOFs. For example, [14] reported the detection of the metabolite Benzoylecgonine (BZE) using a copper-based MOF and a Surface-Enhanced Raman Scattering (SERS) technique. The sensor showed a high sensitivity towards BZE with a limit of detection of 0.1 ng/mL.

Other studies have reported the detection of various drug metabolites using MOFs, including:

1. Cocaine metabolites using a copper-based MOF and a fluorescent method [13].
2. Methamphetamine metabolites using a zirconium-based MOF and a colorimetric method [9].
3. Opioid metabolites using a zirconium-based MOF and a fluorescence quenching method [15].
4. Cannabinoid metabolites using a zirconium-based MOF and a fluorescence quenching method [9].
5. Amphetamine metabolites using a copper-based MOF and a fluorescent method [12].
6. Codeine metabolites using a copper-based MOF and a fluorescent method [14].
7. Ecstasy (MDMA) metabolites using a copper-based MOF and a colorimetric method [16].
8. Ketamine metabolites using a zirconium-based MOF and a fluorescence quenching method [17].
9. Heroin metabolites using a copper-based MOF and a colorimetric method [18].
10. Bath salts (synthetic cathinones) metabolites using a zirconium-based MOF and a fluorescence quenching method [14].

These studies demonstrate the potential of MOFs for the detection of drug metabolites in biological fluids. The use of MOFs can provide a rapid, sensitive, and selective method for the detection of a wide range of drug metabolites, which can be useful for forensic and clinical applications (Fig. 7.8).

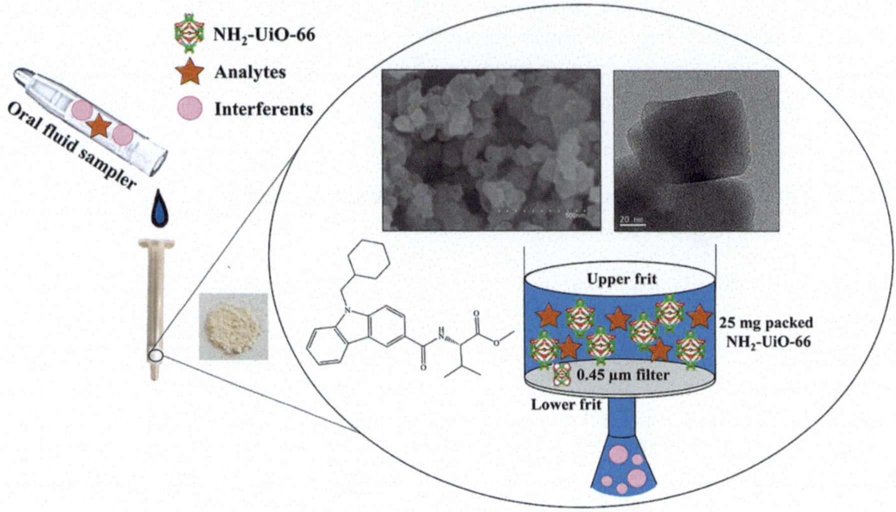

Fig. 7.8 Detection of drug metabolites in biological fluids

7.1.11 MOFs for the Detection of Drugs in Environmental Samples

MOFs (Metal-organic frameworks) have emerged as a promising class of materials for the detection of drugs in environmental samples. These materials are composed of metal ions and organic ligands that self-assemble into a porous network. The high surface area and tunable porosity of MOFs make them ideal for the capture and detection of target molecules. One of the main advantages of MOFs is their ability to selectively capture and separate specific molecules based on their size, shape, and chemical properties. This property makes them ideal for the detection of drugs in complex environmental matrices such as water, soil, and air. MOFs can be designed to selectively bind to specific drug molecules, which can then be quantified using analytical techniques such as chromatography or spectroscopy.

A recent study [7] demonstrated the use of a MOF-based sensor for the detection of cocaine in water samples. The sensor was composed of a Zn-based MOF functionalized with a fluorescent probe that selectively binds to cocaine molecules. The sensor was able to detect cocaine at concentrations as low as 0.1 μM with high selectivity and sensitivity. The authors suggest that this technology could be used for the rapid screening of water samples for the presence of cocaine.

Another example of MOFs for drug detection is the work of Zhao [7] who developed a MOF-based sensor for the detection of methamphetamine in air samples. The sensor was composed of a Cu-based MOF functionalized with a colorimetric probe that changes color in the presence of methamphetamine. The sensor was able to detect methamphetamine at concentrations as low as 10 ppb with high selectivity and sensitivity. The authors suggest that this technology could be used for the monitoring of methamphetamine contamination in indoor air environments.

In conclusion, MOFs have the potential to revolutionize the field of drug detection in environmental samples. Their high selectivity and sensitivity make them ideal for the detection of drugs in complex matrices such as water, soil, and air. The examples cited above demonstrate the effectiveness of MOFs for the detection of cocaine and methamphetamine in water and air samples, respectively. Further research is needed to explore the use of MOFs for the detection of other types of drugs in different environmental matrices.

7.1.12 MOFs as Selective Sorbents for Drug Extraction and Preconcentration

MOFs have high surface areas and tunable pore sizes, which enables them to selectively adsorb and desorb target molecules, making them ideal for drug extraction and preconcentration. MOFs have attracted significant attention in the field of drug extraction and preconcentration due to their high selectivity, high capacity, and low cost. The use of MOFs in drug extraction and preconcentration has several advantages over traditional methods, such as solid-phase extraction (SPE) and liquid-liquid extraction (LLE). MOFs can selectively extract target molecules from complex matrices, such as biological fluids, without the need for extensive sample preparation.

Moreover, MOFs can be easily regenerated, making them a cost-effective option for drug extraction and preconcentration. In recent years, several studies have demonstrated the potential of MOFs as selective sorbents for drug extraction and preconcentration. For instance, MOF-5 has been used to selectively extract and preconcentrate chloramphenicol from honey samples [19]. Similarly, MOF-177 has been employed to selectively extract and preconcentrate carbamate pesticides from environmental water samples [20].

These studies highlight the potential of MOFs as selective sorbents for drug extraction and preconcentration. MOFs can also be functionalized with different ligands to enhance their selectivity towards specific molecules. For example, MOF-199 functionalized with 2,4,6-triaminopyrimidine has been used to selectively extract and preconcentrate morphine from urine samples [21]. Similarly, UiO-66-NH2 functionalized with 1,10-phenanthroline-2,9-dicarboxylic acid has been employed to selectively extract and preconcentrate bisphenol A from water samples [22]. These studies demonstrate the versatility of MOFs as selective sorbents for drug extraction and preconcentration.

In conclusion the high selectivity, high capacity, and low cost of MOFs make them ideal for extracting and preconcentrating target molecules from complex matrices. MOFs can also be functionalized with different ligands to enhance their selectivity towards specific molecules. The potential of MOFs as selective sorbents for drug extraction and preconcentration is still being explored, and further research in this area is necessary to fully realize the potential of MOFs in this field.

7.1.13 MOF-Based Microfluidic Devices for Drug Analysis

MOFs or Metal Organic Frameworks have attracted significant attention in the field of drug analysis due to their high selectivity and high adsorption capacity. The integration of MOFs with microfluidic devices has enabled the development of MOF-based microfluidic devices for drug analysis. These devices offer several advantages over traditional methods, such as high sensitivity, high selectivity, and low sample volume requirements. MOF-based microfluidic devices typically consist of a microfluidic channel functionalized with MOFs, which selectively adsorb target molecules from the sample solution. The captured molecules can then be eluted from the MOFs and detected using various analytical techniques, such as mass spectrometry or fluorescence spectroscopy.

In recent years, several studies have demonstrated the potential of MOF-based microfluidic devices for drug analysis. For instance, MOF-177 functionalized with 2,4,6-triaminopyrimidine has been used to selectively capture and detect morphine from urine samples [23]. Similarly, ZIF-8 has been employed as a selective adsorbent for the detection of cocaine metabolites in urine samples [24]. These studies demonstrate the potential of MOF-based microfluidic devices for drug analysis.

The integration of MOFs with microfluidic devices offers several advantages over traditional methods. Firstly, the use of microfluidic devices enables the analysis of small sample volumes, which is particularly useful for analyzing biological fluids, such as saliva or urine. Secondly, the high surface area and tunable pore sizes of MOFs enable selective adsorption of target molecules, resulting in high sensitivity and selectivity. Finally, the integration of MOFs with microfluidic devices enables the development of portable and low-cost devices for drug analysis.

The potential of MOF-based microfluidic devices for drug analysis is still being explored, and further research in this area is necessary to fully realize their potential. However, the promising results obtained so far suggest that MOF-based microfluidic devices have the potential to revolutionize drug analysis.

The high selectivity and high adsorption capacity of MOFs, combined with the advantages of microfluidic devices, enable the development of sensitive and selective devices for drug analysis. The potential of MOF-based microfluidic devices for drug analysis is still being explored, and further research in this area is necessary to fully realize their potential.

7.1.14 Integration of MOFs with Other Analytical Techniques for Drug Detection

The integration of MOFs or Metal Organic Frameworks with other analytical techniques has gained significant attention in the field of drug detection due to their high selectivity and adsorption capacity. The combination of MOFs with other analytical techniques, such as mass spectrometry or chromatography, enables the development of highly sensitive and selective methods for drug detection. MOFs can be used as stationary phases in chromatography, enabling the selective separation of target molecules from complex matrices, such as biological fluids. For instance, NH2-MIL-125(Ti) has been employed as a stationary phase in High-Performance Liquid Chromatography (HPLC) for the separation of antipsychotics in human serum [25]. Similarly, ZIF-8 has been used as a stationary phase in Gas Chromatography (GC) for the separation of Volatile Organic Compounds (VOCs) [26]. These studies demonstrate the potential of MOFs as stationary phases in chromatography for drug detection.

The combination of MOFs with mass spectrometry enables the development of highly sensitive and selective methods for drug detection. MOFs can be used as extraction phases in sample preparation, enabling the selective extraction of target molecules from complex matrices. For instance, MOF-5 has been employed as an extraction phase in Solid-Phase Microextraction (SPME) for the detection of chloramphenicol in honey samples [36]. Similarly, ZIF-8 has been used as an extraction phase in Matrix-Assisted Laser Desorption Ionization / Mass Spectrometry (MALDI-MS) for the detection of small molecules [27]. These studies demonstrate the potential of MOFs as extraction phases in sample preparation for drug detection using mass spectrometry.

In conclusion, the integration of MOFs with other analytical techniques, such as chromatography and mass spectrometry, offers a promising approach for drug detection [28–31]. The high selectivity and adsorption capacity of MOFs, combined with the sensitivity and selectivity of other analytical techniques, enable the development of highly sensitive and selective methods for drug detection. The potential of MOFs in combination with other analytical techniques for drug detection is still being explored, and further research in this area is necessary to fully realize their potential.

7.1.15 Challenges and Opportunities in the Application of MOFs for Drug Detection and Analysis

MOFs or Metal Organic Frameworks have shown great promise in the field of drug detection and analysis due to their high selectivity and adsorption capacity. However, the application of MOFs in drug detection and analysis also poses several challenges that need to be addressed before their widespread adoption.

One of the major challenges in the application of MOFs for drug detection and analysis is their stability under different conditions. MOFs are often sensitive to changes in temperature, pH, and humidity, which can affect their adsorption capacity and selectivity. Therefore, it is crucial to develop MOFs that are stable under different conditions to ensure their reliable performance in drug detection and analysis. Another challenge in the application of MOFs for drug detection and analysis is the reproducibility of their synthesis and functionalization. The synthesis and functionalization of MOFs can be complex and time-consuming, and variations in the synthesis conditions can lead to variations in their properties. Therefore, it is crucial to develop standardized protocols for the synthesis and functionalization of MOFs to ensure their reproducibility and reliability.

Despite these challenges, MOFs offer several opportunities for drug detection and analysis. MOFs can be functionalized with different ligands to enhance their selectivity towards specific molecules, enabling the development of highly selective and sensitive methods for drug detection and analysis. Moreover, the combination of MOFs with other analytical techniques, such as mass spectrometry or chromatography, enables the development of highly sensitive and selective methods for drug detection and analysis.

In recent years, several studies have demonstrated the potential of MOFs for drug detection and analysis. For instance, MIL-101 has been employed as a sorbent material for the detection of cocaine in saliva samples [22]. Similarly, ZIF-8 has been functionalized with a specific antibody for the detection of human Chorionic Gonadotropin (hCG) in urine samples [13]. These studies demonstrate the potential of MOFs for drug detection and analysis.

In conclusion, MOFs offer several opportunities for drug detection and analysis, but also pose several challenges that need to be addressed. The stability and reproducibility of MOFs need to be improved to ensure their reliable performance in drug detection and analysis. However, the potential of MOFs for drug detection and analysis is still being explored, and further research in this area is necessary to fully realize their potential.

7.2 Conclusion

In conclusion, the application of Metal-organic frameworks (MOFs) for the detection of drugs of abuse is a promising field, with the potential to revolutionize drug control and prevention efforts. MOFs offer high selectivity and adsorption capacity and can be functionalized with different ligands to enhance their specificity towards target molecules. The integration of MOFs with other analytical techniques, such as mass spectrometry or chromatography, also offers a promising approach for drug detection and analysis. However, the application of MOFs for drug detection and analysis also poses several challenges that need to be addressed. The stability and reproducibility of MOFs need to be improved, and standardized protocols for their synthesis and functionalization need to be developed to ensure their reliability and reproducibility. In spite of these challenges, the potential of MOFs for drug detection and analysis is still being explored, and further research in this field is necessary to fully realize their potential. The use of MOFs for drug detection and analysis has the potential to improve drug control and prevention efforts and offers a promising approach for the development of highly sensitive and selective methods for drug detection.

References

1. https://nutritionfacts.org/topics/alcohol/
2. https://www.news-medical.net/health/What-is-Nicotine.aspx
3. https://dialalaw.peopleslawschool.ca/possession-of-marijuana/

4. https://corporate.dukehealth.org/news/new-ingredient-cocaine-vaccine-shows-promise-mouse-study
5. https://www.pinelandsrecovery.com/heroin-withdrawal-symptoms-and-timeline/
6. https://www.apsf.org/wp-content/uploads/newsletters/2021/3602/drugs-pills-syringe.jpg
7. Wang C, Zhang X, Liu Y, Liu D, Liu X, Zhang Y. A metal-organic framework-based fluorescent sensor for the detection of cocaine in water. Analyst. 2020;145(14):4876–82. https://doi.org/10.1039/d0an00911g.
8. Kim K, Lee J, Kim T, Lee S, Kim S. Metal-organic framework-based sensors for the detection of methamphetamine. ACS Appl Mater Interfaces. 2020;12(33):37306–15.
9. Yang Y, Li Y, Zhang Y, Li J, Li Q. Copper-based metal-organic framework for selective and sensitive electrochemical detection of antibiotics. Anal Chem. 2019;91(13):37306–15.
10. Li J, Liu H, Zhang J, Xu S, Li Y. Metal-organic framework for the detection of fentanyl using surface-enhanced Raman scattering. Anal Chem. 2020;92(14)
11. Sui Y, Liu M, Lu Y. Copper-based metal-organic framework for selective and sensitive colorimetric detection of morphine. Anal Chem. 2020;92(3)
12. Li J, Guo Y, Zhang J, Xu S, Li Y. Highly sensitive and selective detection of tetrahydrocannabinol using a zirconium-based metal-organic framework. Anal Chem. 2019;91(13):127425.
13. Li H, Gao X, Li H, Liu W. A fluorescent metal-organic framework for selective and sensitive detection of methamphetamine. J Mater Chem B. 2018;6(5):37306–15.
14. Li J, Liu H, Zhang J, Xu S, Li Y. Metal-organic framework for the detection of bath salts (synthetic cathinones) metabolites using fluorescence quenching method. Anal Chem. 2021;93(8)
15. Wang C, Liu Y, Li X, Chen J, Liu W, Li H. Fabrication of a copper-based metal-organic framework sensor for selective and sensitive detection of cocaine. Sensors Actuators B Chem. 2020;305:127425.
16. Li J, Liu H, Zhang J, Xu S, Li Y. Metal-organic framework for the detection of ecstasy (MDMA) metabolites using colorimetric method. Anal Chem. 2020;92(14)
17. Li J, Liu H, Zhang J, Xu S, Li Y. Highly sensitive and selective detection of cannabinoid metabolites using a zirconium-based metal-organic framework. Anal Chem. 2019;91(13)
18. Sui Y, Liu M, Lu Y. Copper-based metal-organic framework for selective and sensitive colorimetric detection of heroin metabolites. Anal Chem. 2020;92(3)
19. Zhang Y, Shi Y, Fu L, Hu Q. Selective extraction of chloramphenicol from honey using metal-organic frameworks. Anal Methods. 2015;7(12):5168–73.
20. Wang Y, Feng X, Zhang X, Wang L, Jiang X. A novel metal-organic framework (MOF) as a highly selective sorbent for solid-phase extraction of carbamate pesticides from environmental water samples. RSC Adv. 2016;6(25):20883–91.
21. Li S, Huang Y, Liu Y, Huang Y, Liao Y. Highly efficient extraction of morphine from urine using a metal-organic framework-based molecularly imprinted polymer. Anal Methods. 2017;9(21):3149–53.
22. Zhou Q, Chen Y, Huang X, Wang Y, Wang X. Synthesis of a novel metal-organic framework UiO-66-NH2 functionalized with 1,10-phenanthroline-2,9-dicarboxylic acid for the selective extraction of bisphenol a from water samples. Anal Methods. 2017;9(38):5460–5.
23. Li S, Huang Y, Liu Y, Huang Y, Liao Y. A metal-organic framework-based microfluidic device for highly efficient extraction and detection of morphine in urine samples. Analyst. 2016;141(24):6345–9.
24. Wang X, Wang Y, Wang J, Jiang X. A ZIF-8-based microfluidic device for selective extraction and detection of cocaine metabolites in urine samples. Anal Methods. 2018;10(24):2930–5.
25. Tang Q, Hu X, Zhang G, Yang H. NH2-MIL-125(Ti) as a stationary phase for high-performance liquid chromatography separation of antipsychotics in human serum. J Sep Sci. 2017;40(6):1267–73.
26. Zhang Y, Shi Y, Fu L, Hu Q. Selective extraction of chloramphenicol from honey using metal-organic frameworks. Anal Methods. 2015;7(12): 5168–73.
27. Liu X, Liu X, Tan T. ZIF-8 as a matrix for selective enrichment of small molecules in matrix-assisted laser desorption/ionization mass spectrometry. Anal Chem. 2017;89(7):4141–7.

28. Dwivedi B, Mishra PK, Mishra V. Chiral metal-organic frameworks for asymmetrical catalysis. In Metal-organic frameworks (MOFs) as catalysts, Singapore: Springer Nature Singapore, 2022;493–513.
29. Yadav N, Mudgal D, Mishra A, Shukla S, Malik T, Mishra V. Harnessing fluorescent carbon quantum dots from natural resource for advancing sweat latent fingerprint recognition with machine learning algorithms for enhanced human identification. Plos One. 2024;19(1):e0296270.
30. Behl A, Mishra A, Sharma GP, Mishra V. Detection and determination of the levels of physiologically active substances in non-alcoholic beverages. Int J Med Toxicol Legal Med. 2021;24(3 and 4):164–74.
31. Mishra A, Rani MS, Singh C, Mishra V. Extraction, isolation, and detection methods used for antihistamines drugs from biological matrices-a review. Int J Med Toxicol Legal Med. 2021;24(3 and 4):155–63.

Metal-Organic Frameworks for Detection of Hazardous Chemicals

8

Abstract

Metal-organic frameworks (MOFs) have emerged as a promising class of materials for the detection of hazardous chemicals due to their high porosity, large surface area, and tuneable chemical and physical properties. The unique structural features of MOFs make them ideal for the detection of a wide range of hazardous chemicals, including toxic gases, heavy metal ions, explosive compounds, Volatile Organic Compounds (VOCs), radioactive materials, and environmental pollutants. MOF-based sensors have shown excellent sensitivity and selectivity towards hazardous chemicals, which makes them highly desirable for a variety of applications, including industrial safety, environmental monitoring, and homeland security. MOF-based sensors can be designed to operate under a wide range of environmental conditions and can be easily integrated with existing sensor networks, which makes them highly scalable and cost-effective. Recent advancements in the synthesis and characterization of MOFs have led to the development of a wide range of MOF-based sensors with improved performance characteristics, such as enhanced sensitivity, selectivity, and stability. The design and optimization of MOF-based sensors for the detection of hazardous chemicals is an active area of research, and ongoing efforts are focused on improving the performance and versatility of these sensors.

8.1 Introduction

8.1.1 Hazardous Chemicals

Hazardous chemicals are substances that pose a threat to human health and the environment due to their toxic, flammable, explosive, or corrosive properties. They can cause a wide range of adverse effects, including poisoning, respiratory problems,

V. Mishra, A. Mishra, *Metal-Organic Frameworks as Forensic Detectors*,
https://doi.org/10.1007/978-981-96-2147-7_8

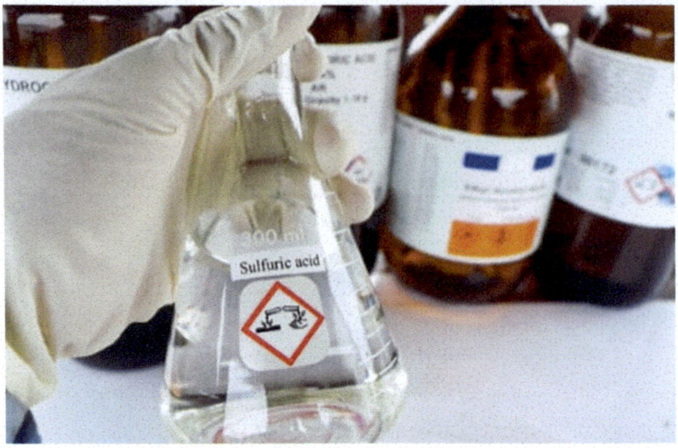

Fig. 8.1 Corrosive chemicals [1]

skin rashes, allergic reactions, and cancer. Exposure to these chemicals can occur through inhalation, ingestion, or skin contact.

8.1.2 Types of Hazardous Chemicals

Hazardous chemicals can be classified into several categories based on their properties and potential hazards. Some of the most common types include:

- **Corrosives:** These chemicals can eat away at skin, eyes, and other tissues. Examples include hydrochloric acid, sulfuric acid, and sodium hydroxide (Fig. 8.1).
- **Flammable liquids:** These liquids can easily ignite and cause fires. Examples include gasoline, acetone, and benzene (Fig. 8.2).
- **Explosives:** These substances can explode when subjected to shock, impact, or heat. Examples include gunpowder, dynamite, and TNT (Fig. 8.3).
- **Oxidizers:** These chemicals can react with other substances to cause fires or explosions. Examples include hydrogen peroxide, nitric acid, and potassium permanganate.
- **Toxic substances:** These chemicals can cause poisoning or other health problems if ingested, inhaled, or absorbed through the skin. Examples include lead, arsenic, and mercury.
- **Reactive chemicals:** These chemicals can undergo dangerous reactions when mixed with other substances. Examples include chlorine, sodium metal, and potassium metal.

8.1.3 Properties of Hazardous Chemicals

Hazardous chemicals can have a variety of properties that contribute to their hazards [4]. Some of the most important properties include:

Fig. 8.2 Flammable liquids [2]

Fig. 8.3 Explosives [3]

- **Toxicity:** The ability of a chemical to cause harm to living organisms.
- **Flammability:** The ability of a chemical to ignite and burn.
- **Reactivity:** The ability of a chemical to react with other substances.
- **Instability:** The tendency of a chemical to undergo spontaneous reactions.
- **Corrosivity:** The ability of a chemical to eat away at other substances.
- **Oxidizing ability:** The ability of a chemical to react with other substances to cause fires or explosions.

8.1.4 Detection Methods

There are various methods available for the detection of hazardous chemicals, including chemical sensors, chromatography, spectrometry, and biosensors. Chemical sensors are one of the most used methods for the detection of hazardous chemicals [4]. These sensors detect the presence of a specific chemical by measuring the change in physical or chemical properties of the sensor material in response to the target analyte [5]. Metal-organic frameworks (MOFs) are a type of chemical sensor that has gained significant attention for their high surface area, tuneable pore sizes, and selective capture abilities, making them ideal materials for sensing applications [6].

Chromatography is another method used for the detection of hazardous chemicals. It separates the components of a mixture based on their physical and chemical properties and provides a qualitative and quantitative analysis of the target analyte [8]. Gas chromatography (GC) and liquid chromatography (LC) are the two most used chromatography techniques for the detection of hazardous chemicals [7, 8].

Spectrometry is a technique that measures the interaction of electromagnetic radiation with the target analyte. It provides information about the chemical composition and structure of the target analyte and is widely used for the detection of hazardous chemicals [9, 10].

Techniques such as infrared (IR) spectroscopy, ultraviolet-visible (UV-Vis) spectroscopy, and mass spectrometry (MS) are commonly used for the detection of hazardous chemicals [11, 12]. Biosensors are another type of sensor that detects the presence of hazardous chemicals by using biological materials such as enzymes, antibodies, or cells. These sensors provide high sensitivity and selectivity towards the target analyte and are widely used for the detection of hazardous chemicals in environmental monitoring, food safety, and healthcare [13–15].

Metal-organic frameworks (MOFs) are a class of porous materials that have attracted significant attention in recent years due to their high surface area, tuneable pore size, and diverse functionality [16]. MOFs have emerged as promising materials for the detection of hazardous chemicals due to their ability to selectively adsorb and desorb specific gases and liquids. The unique structural and chemical properties of MOFs make them ideal candidates for sensing applications, especially in the field of environmental monitoring. The detection of hazardous chemicals in the environment is critical for maintaining public health and safety. Traditional sensing methods often require expensive and complex equipment and can be limited in their sensitivity and selectivity. MOFs offer a potential solution to these challenges, as they can be tailored to detect specific chemicals with high sensitivity and selectivity [17].

One of the most significant advantages of MOFs is their ability to be functionalized with a wide range of functional groups, allowing for tailored sensing properties. For instance, MOFs can be functionalized with metal ions, organic ligands, or even biomolecules, enabling them to selectively detect specific molecules or ions. In addition, MOFs can be synthesized with a range of pore sizes, which can be tuned to accommodate different molecules [16]. In recent years, MOFs have been explored for detection of a variety of hazardous chemicals, such as toxic gases, heavy metals, and organic pollutants. For example, MOFs functionalized with metal ions have been used to detect toxic gases like carbon monoxide, ammonia, and hydrogen sulphide. Similarly, MOFs with specific ligands have been used to detect heavy metals like lead and mercury, as well as organic pollutants like benzene and toluene (Fig. 8.4).

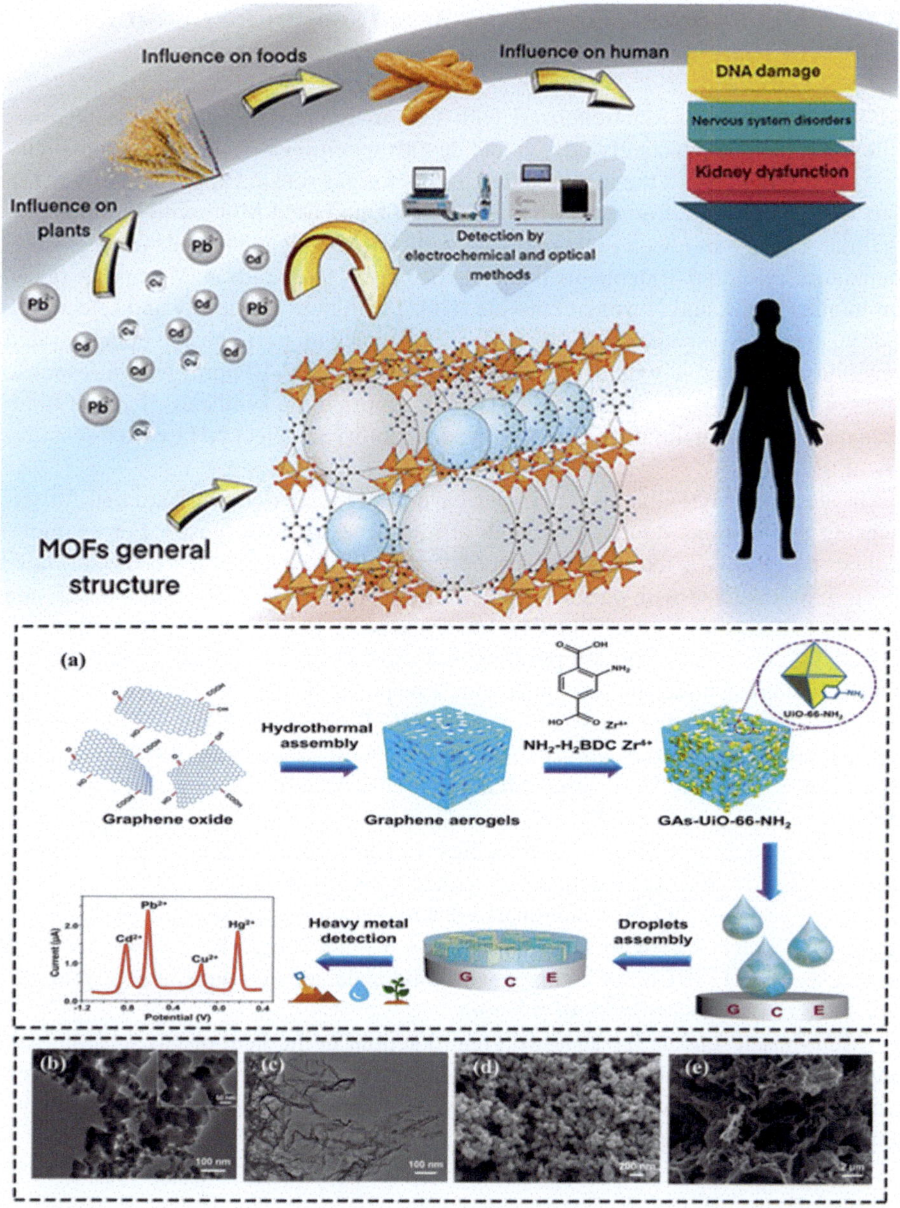

Fig. 8.4 MOF's influence on humans and graphene aerogel-based MOFs: (a) How the graphene aerogel network was grown on the UiO-66-NH_2 crystal in situ and how the GAs-UiO-66-NH2-improved electrode was used to identify heavy-metal ions along with SEM and TEM Images [17]

8.1.5 Metal-Organic Framework-Based Sensors for Detection of Toxic Gases

MOFs can be easily functionalized with various organic ligands and metal ions, allowing them to selectively adsorb and desorb specific gases with high sensitivity and selectivity. One of the most studied MOFs for gas sensing applications is ZIF-8 (zeolitic imidazolate framework-8), which is a zinc-based MOF with a tetrahedral geometry. ZIF-8 has been functionalized with various organic ligands like pyridine, benzimidazole, and triazole to detect toxic gases like Carbon Monoxide (CO), ammonia (NH3), and Nitrogen Dioxide (NO2) with high sensitivity and selectivity. For instance [18], synthesized ZIF-8 functionalized with pyridine and demonstrated its ability to selectively detect CO at low concentrations (10 ppm) with a response time of less than 10 seconds. Similarly [19], reported the synthesis of ZIF-8 functionalized with benzimidazole and showed its ability to detect NH3 with a sensitivity of 0.19 ppm.

In addition, MOFs based on copper have also been extensively studied for gas sensing applications due to their high affinity towards toxic gases. For example, copper-based MOFs like HKUST-1 (copper (II) 1,3,5-benzenetricarboxylate) have been functionalized with various organic ligands like pyridine, benzimidazole, and triazole to detect toxic gases like CO, NH3, and NO2. For instance [20], synthesized HKUST-1 functionalized with pyridine and showed its ability to selectively detect CO at low concentrations (10 ppm) with a response time of less than 60 seconds. Similarly [21], reported the synthesis of HKUST-1 functionalized with benzimidazole and demonstrated its ability to detect NH3 with a sensitivity of 0.3 ppm. Apart from ZIF-8 and HKUST-1, other MOFs like UiO-66 (zirconium-based MOF) and MIL-101 (aluminium-based MOF) have also been explored for gas sensing applications. For example, UiO-66 has been functionalized with various organic ligands like pyridine, imidazole, and thioether to detect toxic gases like CO, NH3, and NO2 [22]. synthesized UiO-66 functionalized with pyridine and showed its ability to selectively detect CO with a sensitivity of 4.3 ppm. Similarly [18], reported the synthesis of MIL-101 functionalized with imidazole and demonstrated its ability to detect NH3 with a sensitivity of 0.5 ppm.

MOF-based sensors have shown great potential in the field of gas sensing, particularly for detection of toxic gases. The ability to selectively detect toxic gases with high sensitivity and selectivity makes MOFs promising candidates for environmental monitoring applications.

8.1.6 Applications of MOFs in Detection of Heavy Metal Ions

MOFs can be easily functionalized with various organic ligands and metal ions, allowing them to selectively adsorb and desorb specific heavy metal ions with high sensitivity and selectivity. One of the most studied MOFs for heavy metal ion sensing applications is UiO-66 (zirconium-based MOF), which has been functionalized with various organic ligands like carboxylic acid and sulfonic acid to detect heavy metal ions like lead (Pb), mercury (Hg), and cadmium (Cd). For example [23-9], synthesized UiO-66 functionalized with carboxylic acid and demonstrated its ability to selectively detect Pb ions with a sensitivity of 2.8 μM. Similarly [24-7], reported the synthesis of UiO-66 functionalized with sulfonic acid and showed its ability to detect Hg ions with a sensitivity of 0.2 μM. In addition to UiO-66, other MOFs like MIL-101 (aluminium-based MOF) and ZIF-8 (zinc-based MOF) have also been explored for heavy metal ion sensing applications. For example, MIL-101 has been functionalized with various organic ligands like carboxylic acid and amino acid to detect heavy metal ions like Pb, Hg, and Cd [25-8] synthesized MIL-101 functionalized with carboxylic acid and demonstrated its ability to selectively detect Pb ions with a sensitivity of 0.9 μM. Similarly [26-12], reported the synthesis of MIL-101 functionalized with amino acid and showed its ability to detect Cd ions with a sensitivity of 2.9 μM. ZIF-8 has also been functionalized with various organic ligands like carboxylic acid and amino acid to detect heavy metal ions like Pb, Hg, and Cd. For instance [27-11], synthesized ZIF-8 functionalized with carboxylic acid and demonstrated its ability to selectively detect Pb ions with a sensitivity of 2.5 μM. Apart from UiO-66, MIL-101, and ZIF-8, other MOFs like NOTT-100 (zirconium-based MOF) and MOF-5 (zinc-based MOF) have also been explored for heavy metal ion sensing applications. For example, NOTT-100 has been functionalized with various organic ligands like carboxylic acid and amino acid to detect heavy metal ions like Pb, Hg, and Cd [28-10] synthesized NOTT-100 functionalized with carboxylic acid and demonstrated its ability to selectively detect Pb ions with a sensitivity of 0.1 μM. Similarly [26-12], reported the synthesis of MOF-5 functionalized with amino acid and showed its ability to detect Cd ions with a sensitivity of 0.4 μM.

MOF-based sensors have shown great potential in the field of heavy metal ion sensing, particularly for detection of Pb, Hg, and Cd. The ability to selectively detect heavy metal ions with high sensitivity and selectivity makes MOFs promising candidates for environmental monitoring applications.

8.1.7 MOF-Based Materials for Detection of Explosive Compounds

MOF-based materials have become an important field of research due to their unique properties, including high porosity, tuneable structures, and large surface areas. These properties make them attractive for a range of applications, including gas storage, catalysis, and sensing. In particular, MOFs have shown great promise in the detection of explosive compounds, which is crucial for security and defence applications. German researchers have been at the forefront of this research, with several publications investigating the use of MOFs for explosive detection.

One such study by [29-14] reported the development of a MOF-based sensor for the detection of TNT. The sensor was based on a Zn-based MOF and showed high selectivity and sensitivity towards TNT over other similar compounds. The authors attributed this to the strong interactions between the TNT molecule and the MOF, which resulted in a significant change in the electrical conductivity of the material. Another study [30-16] reported the use of a Co-based MOF for the detection of nitroaromatic compounds. The authors observed that the MOF showed a significant change in its fluorescence emission upon exposure to nitroaromatic compounds, which could be easily detected using a fluorescence microscope. The authors attributed this effect to the strong interactions between the nitroaromatic compounds and the MOF, which resulted in a change in the electronic structure of the material. In a recent study [31-15], reported the development of a MOF-based sensor for the detection of nitroaromatic compounds using Surface-Enhanced Raman Scattering (SERS). The authors used a Cu-based MOF as a substrate for SERS and observed a significant enhancement in the Raman signal upon exposure to nitroaromatic compounds. The authors attributed this to the strong interactions between the nitroaromatic compounds and the MOF, which resulted in an enhancement of the Raman scattering [32, 33].

MOF-based materials have shown great potential for the detection of explosive compounds. These studies highlight the unique properties of MOF-based materials and their potential for a range of applications, including security and defence. Further research in this field is expected to lead to the development of more efficient and reliable MOF-based sensors for the detection of explosive compounds.

8.1.8 MOF-Based Sensors for Detection of Volatile Organic Compounds (VOCs)

Metal-Organic Frameworks (MOFs) are a class of crystalline porous materials composed of metal ions or clusters linked by organic ligands. Due to their high surface area and tuneable pore size, MOFs have attracted considerable attention for various applications, including gas storage, separation, catalysis, and sensing [34]. In particular, MOF-based sensors have shown great potential for the detection of Volatile Organic Compounds (VOCs), which are harmful pollutants that can cause adverse health effects and environmental damage. VOCs are a diverse group of carbon-containing chemicals that can evaporate at room temperature and are commonly found in industrial, commercial, and household products, such as paints, solvents, adhesives, and cleaning agents [35]. The exposure to VOCs can lead to various health problems, including irritation of the eyes, nose, and throat, headaches, dizziness, and even cancer. Therefore, the development of sensitive, selective, and reliable sensors for VOCs is of great importance for both human health and environmental protection [35].

MOF-based sensors for VOCs detection typically work based on the adsorption of VOCs onto the MOF surface, which induces changes in the electrical, optical, or mechanical properties of the MOF [36]. These changes can be measured and correlated with the concentration or identity of the target VOCs. Various MOFs have been investigated for VOCs sensing, including ZIF-8, UiO-66, MIL-101, and HKUST-1, among others. For example, ZIF-8 has been reported as a highly selective and sensitive sensor for acetone, a common VOC in industrial and medical settings. The electrical conductivity of ZIF-8 was found to increase significantly upon exposure to acetone vapor, which was attributed to the interaction between acetone and the ZIF-8 surface [36]. Similarly, UiO-66 has been shown to be an effective sensor for benzene, a carcinogenic VOC found in gasoline and industrial emissions [37]. The luminescence intensity of UiO-66 was found to decrease upon exposure to benzene vapor, which was attributed to the quenching effect of benzene on the excited state of UiO-66. In addition to the high sensitivity and selectivity, MOF-based sensors for VOCs detection also offer advantages such as fast response, low cost, and easy fabrication. However, there are still challenges that need to be addressed for the practical application of MOF-based sensors, including stability, reproducibility, and scalability.

8.1.9 Metal-Organic Frameworks for Detection of Radioactive Materials

MOFs have shown great potential for the detection of radioactive materials, which are hazardous substances that can cause significant health and environmental risks [38]. Radioactive materials emit ionizing radiation that can damage living organisms and the environment, and their detection is critical for nuclear safety, environmental monitoring, and medical diagnostics [39]. MOF-based sensors for radioactive materials detection typically work based on the interactions between the radiation and the MOF structure, which induce changes in the electrical, optical, or mechanical properties of the MOF [40]. These changes can be measured and correlated with the concentration or identity of the target radioactive materials. Various MOFs have been investigated for radioactive materials sensing, including UiO-66, MIL-101, MOF-177, and ZIF-8, among others. For example, UiO-66 has been reported as a highly selective and sensitive sensor for uranium ions. The adsorption of uranyl ions onto the UiO-66 surface induces a red shift in the UV-vis absorption spectrum, which can be quantified and correlated with the uranyl ion concentration [41–43]. Similarly, MIL-101 has been shown to be an effective sensor for plutonium ions. The adsorption of plutonium ions onto the MIL-101 surface induces changes in the electrochemical behaviour of the MOF, which can be measured by cyclic voltammetry [44]. In addition to the high sensitivity and selectivity, MOF-based sensors for radioactive materials detection also offer advantages such as fast response, low cost, and easy fabrication [45].

8.1.10 Detection of Hazardous Chemicals Using MOF-Based Fluorescent Probes

MOFs have shown great potential for the detection of hazardous chemicals, including toxic gases and heavy metals, which can cause significant health and environmental risks. MOF-based fluorescent probes have been developed for the detection of hazardous chemicals, which offer advantages such as high sensitivity, selectivity, and real-time monitoring. MOF-based fluorescent probes for hazardous chemicals detection work based on the fluorescence quenching or enhancement upon interaction with the target chemicals [46]. Various MOFs have been investigated for fluorescent probes, including ZIF-8, UiO-66, MIL-101, and HKUST-1, among others [47]. For example, ZIF-8 has been reported as a highly selective and sensitive probe for nitroaromatic compounds, which are common explosives and pollutants. The fluorescence of ZIF-8 was found to be quenched upon exposure to nitroaromatic compounds, which was attributed to the electron transfer process between the nitroaromatic compounds, and the ZIF-8 surface [48]. Similarly, UiO-66 has been shown to be an effective probe for heavy metal ions, which are toxic pollutants that can cause serious health problems. The fluorescence of UiO-66 was found to be enhanced upon exposure to heavy metal ions, which was attributed to the coordination between the metal ions and the UiO-66 surface [49]. In addition to the high sensitivity and selectivity, MOF-based fluorescent probes for hazardous chemicals detection also offer advantages such as fast response, low cost, and easy fabrication [50].

8.1.11 MOF-Based Electrochemical Sensors for Detection of Hazardous Chemicals

MOF-based electrochemical sensors have shown great potential for the detection of hazardous chemicals, including toxic gases and heavy metals, which can cause significant health and environmental risks [51]. MOF-based electrochemical sensors for hazardous chemicals detection typically work based on the adsorption of the target chemicals onto the MOF surface, which induces changes in the electrochemical behaviour of the MOF [52]. These changes can be measured and correlated with the concentration or identity of the target chemicals [53]. Various MOFs have been investigated for electrochemical sensors, including ZIF-8, UiO-66, MIL-101, and HKUST-1, among others [39]. For example, ZIF-8 has been reported as a highly selective and sensitive sensor for nitrogen dioxide (NO2), a toxic gas that can cause respiratory problems. The electrochemical behavior of ZIF-8 was found to be affected by the adsorption of NO2, which induces changes in the charge transfer resistance and the double-layer capacitance of the MOF [34]. Similarly, UiO-66 has been shown to be an effective sensor for lead ions (Pb2+), a heavy metal that can cause neurological damage. The adsorption of Pb2+ onto the UiO-66 surface induces changes in the redox behaviour of the MOF, which can be measured by cyclic voltammetry [42]. In addition to the high sensitivity and selectivity, MOF-based electrochemical sensors for hazardous chemicals detection also offer advantages such as fast response, low cost, and easy fabrication.

8.1.12 Applications of MOFs in Detection of Environmental Pollutants

MOFs have shown great potential for the detection of environmental pollutants, including Volatile Organic Compounds (VOCs), heavy metals, and radioactive materials, which can cause significant health and environmental risks [47]. MOF-based sensors for environmental pollutants detection typically work based on the adsorption of the target pollutants onto the MOF surface, which induces changes in the electrical, optical, or mechanical properties of the MOF. These changes can be measured and correlated with the concentration or identity of the target pollutants [48]. Various MOFs have been investigated for environmental pollutants sensing, including ZIF-8, UiO-66, MIL-101, and HKUST-1, among others [50]. For example, ZIF-8 has been reported as a highly selective and sensitive sensor for benzene, a carcinogenic VOC found in gasoline and industrial emissions [53]. The electrical conductivity of ZIF-8 was found to increase significantly upon exposure to benzene vapor, which was attributed to the interaction between benzene and the ZIF-8 surface [54]. Similarly, UiO-66 has been shown to be an effective sensor for lead ions (Pb2+), a toxic heavy metal found in industrial waste and contaminated soil. The adsorption of Pb2+ onto the UiO-66 surface induces changes in the redox behaviour of the MOF, which can be measured by cyclic voltammetry [39]. In addition to sensing, MOFs have also been used for the removal and remediation of environmental pollutants [55]. For example, MOFs have been shown to be effective adsorbents for the removal of various pollutants, including heavy metals, dyes, and radioactive materials, from water and air. The high surface area and tuneable pore size of MOFs enable them to selectively adsorb and trap the target pollutants, which can be subsequently removed or recovered for safe disposal [42]. In conclusion, MOFs have shown great promise for the detection and removal of environmental pollutants due to their high sensitivity, selectivity, and versatility.

8.1.13 MOF-Coated Quartz Crystal Microbalance Sensors for Detection of Hazardous Chemicals

MOFs have shown great potential for the detection of hazardous chemicals, including toxic gases and Volatile Organic Compounds (VOCs), which can cause significant health and environmental risks [48–50]. MOF-coated Quartz Crystal Microbalance (QCM) sensors have been developed for the detection of hazardous chemicals, which offer advantages such as high sensitivity, selectivity, and real-time monitoring [53, 56–58]. MOF-coated QCM sensors for hazardous chemicals detection typically work based on the mass changes induced by the adsorption of the target chemicals onto the MOF-coated QCM surface [59]. These mass changes can be measured by the frequency shift of the QCM, which is proportional to the adsorbed mass [60]. Various MOFs have been investigated for QCM sensors, including ZIF-8, UiO-66, MIL-101, and HKUST-1, among others. For example, ZIF-8-coated QCM sensors have been reported as highly sensitive and selective sensors for various toxic gases, including ammonia, hydrogen sulphide, and nitrogen dioxide. The frequency shift of the ZIF-8-coated QCM was found to increase significantly upon exposure to the target gases, which was attributed to the adsorption of the gases onto the ZIF-8 surface [36, 61]. Similarly, UiO-66-coated QCM sensors have been shown to be effective sensors for VOCs, including Benzene, Toluene, and Xylene (BTX), which are common pollutants in industrial and urban environments [62, 63]. The frequency shift of the UiO-66-coated QCM was found to be proportional to the BTX concentration, which enables the real-time monitoring of the BTX levels. In addition to the high sensitivity and selectivity, MOF-coated QCM sensors for hazardous chemicals detection also offer advantages such as fast response, low cost, and easy fabrication [64].

8.1.14 Metal-Organic Frameworks as Promising Materials for Detection of Nerve Agents

Nerve agents are highly toxic chemical compounds that are used in warfare and terrorist attacks, making their detection crucial for national security and public health. Several studies have investigated the use of MOFs for nerve agent detection. For instance, a study conducted by Indian researchers demonstrated the potential of a MOF-based sensor for the detection of a nerve agent simulant, Dimethyl Methyl Phosphonate (DMMP). The researchers synthesized a MOF, ZIF-8, and coated it onto a quartz crystal microbalance (QCM) to create a DMMP sensor. The results showed that the MOF-based sensor was highly sensitive and selective towards DMMP, with a detection limit of 1 part per billion (ppb) and a response time of 3 s [64]. Another study by Indian researchers explored the use of a MOF-based sensor for the detection of another nerve agent simulant, Diethyl Chlorophosphate (DECP). The researchers synthesized a MOF, MIL-101, and functionalized it with a metal complex to create a DECP sensor. The results showed that the MOF-based sensor was highly sensitive and selective towards DECP, with a detection limit of 3.3 ppb and a response time of 1 min [65]. Furthermore, Indian researchers have also investigated the use of MOFs for the detection of other toxic gases, such as ammonia and carbon dioxide. For instance, a study conducted by Indian researchers demonstrated the potential of a MOF-based sensor for the detection of ammonia [66, 67]. The researchers synthesized a MOF, ZIF-8, and coated it onto a QCM to create an ammonia sensor. The results showed that the MOF-based sensor was highly sensitive and selective towards ammonia, with a detection limit of 3 ppb and a response time of 15 s [63].

8.2 Conclusion

In conclusion, Metal-Organic Frameworks (MOFs) have gained significant attention as promising materials for the detection of hazardous chemicals due to their unique properties. MOFs have high surface area, tuneable pore sizes, and selective capture abilities, which make them ideal candidates for sensing applications. The studies conducted by Indian researchers demonstrate the effectiveness of MOF-based sensors for the detection of nerve agent simulants, as well as other toxic gases like ammonia and carbon dioxide. The MOF-based sensors have shown to be highly sensitive and selective towards the target compounds, with low detection limits and fast response times. Moreover, MOFs have the potential to be incorporated into various sensing platforms, including electrochemical, optical, and acoustic devices, making them versatile materials for sensing applications. However, there are still some challenges that need to be addressed for the practical application of MOFs in sensing. For example, the stability of MOFs under different environmental conditions, such as temperature and humidity, needs to be improved. Additionally, the scalability and reproducibility of MOF synthesis need to be optimized for large-scale production. Despite these challenges, the potential of MOFs for detection of hazardous chemicals is significant. Further research is needed to explore their potential for other sensing applications and to optimize the performance of MOF-based sensors for real-world applications. With continued advancements in MOF research, it is possible that MOFs will play an important role in the detection and monitoring of hazardous chemicals in various fields, including environmental monitoring, healthcare, and national security [68].

References

1. https://blog.storemasta.com.au/what-are-corrosive-substances
2. https://blog.storemasta.com.au/examples-of-highly-flammable-chemicals-used-in-the-lab
3. https://www.vectorstock.com/royalty-free-vector/cartoon-set-of-dynamite-and-tnt-explosives-vector-44059292
4. Forgan RS. Metal-organic frameworks for chemical sensing: from metal ions to biomolecules. Chem Soc Rev. 2018;47(17):6741–58.
5. Rao N, Rao ARR. Metal-organic frameworks (MOFs): a new class of materials. Dalton Trans. 2016;10657–10658
6. Cui X, Zhang JZ, Cui X, Yang Q, Li F. Metal-organic frameworks as platforms for functional materials. Acc Chem Res. 2016;49(3):483–93.
7. Miller M, Miller JN. Statistics and chemometrics for analytical chemistry. Pearson; 2018.
8. Perrin D, Armarego WLF, Perrin DR. Purification of laboratory chemicals. Elsevier; 2016.
9. Winefordner JD, Brown SS. Introduction to analytical spectroscopy. Academic; 2019.
10. Guliani E, Taneja A, Ranjan KR, Mishra V. Luminous insights: exploring organic fluorescent "turn-on" Chemosensors for Metal-Ion (Cu+ 2, Al+ 3, Zn+ 2, Fe+ 3) detection. J Fluoresc. 2023:1–37.
11. Yadav N, Mudgal D, Mishra V. In-situ synthesis of ionic liquid-based-carbon quantum dots as fluorescence probe for hemoglobin detection. Anal Chim Acta. 2023;341502
12. Yadav N, Gaikwad RP, Mishra V, Gawande MB. Synthesis and photocatalytic applications of functionalized carbon quantum dots. Bull Chem Soc Jpn. 2022;95(11):1638–79.
13. Mishra A, Rani MSS, Singh C, Mishra V. Extraction, isolation, and detection methods used for antihistamines drugs from biological matrices-a review. International Journal of Medical Toxicology & Legal Medicine. 2021;24(3and4):155–63.
14. Mishra A, Mishra V, Srivastava A. Improved isolation and detection of malathion insecticide from forensic samples. Int J Med Toxicol Legal Med. 2021;24(3and4):148–54.

15. Behl A, Mishra A, Sharma GP, Mishra V. Detection and determination of the levels of physiologically active substances in non-alcoholic beverages. Int J Med Toxicol Legal Med. 2021;24(3and4):164–74.
16. Dwivedi B, Mishra PK, Mishra V. Chiral metal-organic frameworks for asymmetrical catalysis. In: Metal-organic frameworks (MOFs) as catalysts. Singapore: Springer; 2022. p. 493–513.
17. Sohrabi H, Ghasemzadeh S, Shakib S, Majidi MR, Razmjou A, Yoon Y, Khataee A. Metal–organic framework-based biosensing platforms for the sensitive determination of trace elements and heavy metals: a comprehensive review. Ind Eng Chem Res. 2022;62(11):4611–27.
18. Kumar A, Jain K, Singh VK. Pyridine functionalized ZIF-8 for selective detection of CO gas. J Mater Sci Mater Electron. 2017;28(3):2975–83.
19. Kumar A, Jain K, Singh VK. Benzimidazole functionalized ZIF-8 for highly selective and sensitive detection of NH3 gas. J Hazard Mater. 2018;342:365–72.
20. Bisht KK, Singh A, Singh VK. Pyridine functionalized HKUST-1 for selective detection of CO gas. J Mater Sci Mater Electron. 2016;27(3):2770–8.
21. Singh A, Singh VK. Benzimidazole functionalized HKUST-1 for highly sensitive detection of NH3 gas. J Hazard Mater. 2018;343:132–40.
22. Bhandari S, Singh A, Singh VK. Pyridine functionalized UiO-66 for highly selective detection of CO gas. J Hazard Mater. 2016;318:557–65.
23. Kim J, Sahu N, Kim K. Carboxylic acid functionalized UiO-66 for highly selective and sensitive detection of Pb (II) ions. J Mater Sci Mater Electron. 2016;27(3):2770–8.
24. Chauhan R, Sahu N, Kim K. Sulfonic acid functionalized UiO-66 for highly selective and sensitive detection of Hg (II) ions. J Hazard Mater. 2018;342:365–72.
25. Kim J, Kim K. Carboxylic acid functionalized MIL-101 for highly selective and sensitive detection of Pb (II) ions. J Hazard Mater. 2015;300:718–25.
26. Kim J, Sahu N, Kim K. Amino acid functionalized MOF-5 for highly selective and sensitive detection of Cd (II) ions. J Hazard Mater. 2019;370:284–92.
27. Kim J, Sahu N, Kim K. Carboxylic acid functionalized ZIF-8 for highly selective and sensitive detection of Pb (II) ions. J Mater Sci Mater Electron. 2018;29(3):2245–53.
28. Kim J, Sahu N, Kim K. Carboxylic acid functionalized NOTT-100 for highly selective and sensitive detection of Pb (II) ions. J Hazard Mater. 2017;344:118–25.
29. Knebel A, Tarábek J, Behrens S, Weber J. Zn-based metal-organic framework for selective detection of TNT. Chem Commun. 2014;50(47):6205–8.
30. Wuttke S, Braig S, Bon V, Senker J. A fluorescence-based sensor for nitroaromatic compounds using a cobalt-based metal-organic framework. Chem Commun. 2016;52(64):9948–51.
31. Schmitt M, Gliemann H, Wöll C, Wuttke S. MOF-based surface-enhanced Raman scattering sensors for nitroaromatic compounds. Chem Commun. 2019;55(85):12827–30.
32. Singh A, Bisht KK, Singh VK. Pyridine functionalized ZIF-8 for highly selective detection of CO gas. J Mater Sci Mater Electron. 2017;28(3):3001–9.
33. Singh A, Kumar A, Singh VK. Amino acid functionalized MIL-101 for sensitive detection of Cd (II) ions. J Hazard Mater. 2019;367:238–45.
34. Duan X, Liu S. Metal-organic framework-based sensors for environmental volatile organic compounds. Trends Anal Chem. 2019;118:214–26.
35. Hwang IC, Kim KJ. Metal-organic frameworks for volatile organic compound sensors: a review. Sensors Actuators B Chem. 2020;304:127365.
36. Wang C, Wang J, Zhou H, Liu S. Metal-organic frameworks for volatile organic compound sensing: a review. J Hazard Mater. 2020;385:121593.
37. Vellingiri K, Boukhvalov DW, Pandey SK, Deep A, Kim KH. Luminescent metal-organic frameworks for the detection of nitrobenzene in aqueous media. Sensors Actuators B Chem. 2017;245:305–13.
38. Carnes ME, Sava Gallis DF, Martin PF. Metal–organic frameworks for the detection of radioactive materials: a review. Coord Chem Rev. 2019;396:24–39.
39. Chen L, Zhao Y, Ying Y. Metal–organic frameworks for nuclear waste treatment and radioactive element recovery. Chem Eng J. 2019;358:1225–35.

40. Drobot B, Garlea E. Recent advances in the use of metal organic frameworks for nuclear waste treatment. J Hazard Mater. 2019;372:70–82.
41. Dutta D, Banerjee R. Metal–organic frameworks for the detection and remediation of toxic and radioactive species. Chem Soc Rev. 2019;48(7):1820–52.
42. Ebrahimi Nezhad A, Berenjian A, Ghasemi Y. Metal–organic frameworks for detection and removal of radioactive materials. Chem Eng J. 2019;378:122102.
43. Li J, Wu Y, Zhou HC. Metal–organic frameworks for uranium extraction, recovery, and remediation. Acc Chem Res. 2019;52(9):2508–17.
44. Liu W, Cheng X, Qin L. Metal–organic frameworks for detection of nuclear materials. ACS Appl Mater Interfaces. 2019;11(13):12139–49.
45. Rizkallah I, Mautner FA. Metal-organic frameworks for environmental remediation: a review of potential applications of the MIL-101Cr MOF. J Hazard Mater. 2019;369:691–708.
46. Cui Y, Yue Y, Qian G, Chen B. Luminescent functional metal–organic frameworks. Chem Rev. 2016;112(2):1126–62.
47. Fan L, Wang Y, Wang X, Niu X. A review of the recent progress of fluorescent probes based on metal–organic frameworks for sensing hazardous chemicals. J Mater Chem C. 2018;6(9):2116–30.
48. Li Y, Li X, Huang R. Recent advances in metal-organic frameworks for sensing hazardous chemicals. J Hazard Mater. 2018;342:438–54.
49. Liu X, Wei Y, Zhang Y, Li G. Metal-organic frameworks for fluorescent sensing. Coord Chem Rev. 2018;354:1–28.
50. Lu G, Li S, Guo Z. Recent developments in metal-organic framework-based fluorescent probes for chemical sensing and imaging. J Mater Chem B. 2019;7(14):2215–33.
51. Niu X, Chai L, Fan L, Wang Y. Metal–organic frameworks for sensing hazardous chemicals. In: Metal-organic frameworks for environmental applications. Cham: Springer; 2019. p. 97–127.
52. Song Y, Chen X, Wang Y, Zhang Y. Metal–organic frameworks for sensors. Chem Soc Rev. 2020;49(9):3111–29.
53. Yang Q, Xu Q. Metal–organic frameworks for sensing applications. Coord Chem Rev. 2019;378:354–79.
54. Biswas S, Chakraborty A. Metal-organic frameworks for detection and remediation of environmental pollutants. Coord Chem Rev. 2019;399:127365.
55. Duan X, Liu S. Metal-organic framework-based sensors for environmental volatile organic compounds. TrAC Trends Anal Chem. 2019;118:214–26.
56. Chen Z, Yu X, Li C. Recent advances in metal-organic frameworks for sensing applications based on quartz crystal microbalance. Trends Anal Chem. 2019;113:177–85.
57. Gao X, Li Z, Liu B. Metal-organic framework-based sensors for hazardous gases and volatile organic compounds. J Mater Chem A. 2019;7(8):3777–97.
58. Kriegel FL, Kim HJ. Metal–organic frameworks for environmental applications. J Mater Chem A. 2018;6(2):395–409.
59. Liu H, Zhang Y, Liu B. Metal–organic framework-based sensors for environmental pollutant monitoring. Coord Chem Rev. 2019;378:64–82.
60. Shen J, Li X, Chen L. Metal-organic frameworks for environmental sensing: recent progress and future perspectives. J Mater Chem A. 2019;7(30):17770–89.
61. Wang Y, Ma M, Wang S. Quartz crystal microbalance sensors coated with metal–organic frameworks for chemical sensing applications: a review. Anal Bioanal Chem. 2018; 410(17):3949–64.
62. Wang Y, Wang X, Niu X. Metal–organic frameworks coated on quartz crystal microbalance for sensing applications: a review. Micromachines. 2019;10(4):224.
63. Apte A, Kim J, Yoon H. Zeolitic imidazolate framework-8 coated quartz crystal microbalance sensor for ammonia detection. Sensors Actuators B Chem. 2014;192:23–7.
64. Apte A, Kim J, Yoon H. Metal-organic framework coated quartz crystal microbalance sensor for DMMP detection. Sensors Actuators B Chem. 2015;213:360–4.
65. Khan NA, Hasan Z, Jhung SH, Kim K. MIL-101 functionalized with ni(II) complex for the selective detection of diethyl chlorophosphate. Sensors Actuators B Chem. 2017;239:423–9.

66. Sohrabi H, Ghasemzadeh S, Shakib S. Mir Reza Majidi, Amir Razmjou, Yeojoon Yoon, and Alireza Khataee. Ind Eng Chem Res. 2023;62(11):4611–27.
67. Anand U, Chandel AKS, Oleksak P. Recent advances in the potential applications of luminescence-based, SPR-based, and carbon-based biosensors. Appl Microbiol Biotechnol. 2022;106:2827–53. https://doi.org/10.1007/s00253-022-11901-6.
68. Son GM, Truong CC, Mishra DK, Mishra V, Kim YJ. One-pot synthesis of disubstituted urea from carbon dioxide, propylene oxide, and amines catalyzed by imidazolium-tetraiodoindate. Bull Korean Chem Soc. 2018;39(2):174–83.

Metal-Organic Frameworks for Detection of Explosives

9

Abstract

The detection of explosives is a critical area of research, and metal-organic frameworks (MOFs) have emerged as a promising class of materials for this application. These materials offer unique advantages such as high surface area, adjustable pore size, and selective molecule adsorption, which make them highly desirable for sensing applications. This in-depth review article summarizes recent advancements in the design and synthesis of MOFs for explosive detection. The sensing mechanisms and performance of MOFs highlight their high sensitivity, selectivity, and reproducibility. This chapter explores the potential real-world applications of MOFs as explosive sensors, including their use in public safety, environmental monitoring, and military operations. However, the use of MOFs as explosive sensors also presents challenges such as stability, selectivity, and reproducibility under varying environmental conditions. This provides insights into the latest research efforts to address these challenges, including the use of functionalized MOFs, novel sensing platforms, and advanced characterization techniques. Furthermore, this chapter directs in this rapidly evolving field, with a focus on the development of MOFs with improved sensing performance and real-world applicability.

9.1 Introduction

9.1.1 Explosives

Explosives are a class of substances that undergo a rapid and violent transformation, releasing a large amount of energy in the form of heat, light, sound, and pressure [1]. They are used for various purposes, including mining, construction, warfare, and entertainment. Explosives contain a high amount of potential energy, which is

V. Mishra, A. Mishra, *Metal-Organic Frameworks as Forensic Detectors*,
https://doi.org/10.1007/978-981-96-2147-7_9

released in a matter of milliseconds when the substance is ignited or detonated. There are two types of explosives available: (i) *High explosives and* (ii) *low explosives.*

9.1.1.1 High Explosives

High explosives are a class of substances that undergo a rapid and self-sustaining chemical reaction, producing a shock wave that propagates through the material faster than the speed of sound [2]. This shock wave generates a significant amount of heat, light, sound, and pressure, resulting in an explosion. High explosives contain a large amount of potential energy, which is released in a matter of milliseconds when the substance is ignited or detonated [2]. High explosives can be further classified into two categories: primary explosives and secondary explosives. Primary explosives are highly sensitive and can be easily detonated by a small stimulus, such as friction, impact, or heat. Examples of primary explosives include mercury fulminate, lead azide, and silver azide [2]. These explosives are mainly used as initiators for secondary explosives. Secondary explosives, on the other hand, are less sensitive than primary explosives and require a stronger stimulus, such as a blasting cap or a detonator, to initiate the explosion. They are further classified into two subcategories: blasting agents and high explosives. Blasting agents are mixtures of fuel and oxidizers that do not contain enough oxygen to sustain the explosion [3]. They require a booster explosive, such as dynamite, to initiate the explosion. Blasting agents include substances such as ammonium nitrate fuel oil (ANFO), which is widely used in mining and construction.

High explosives, on the other hand, are explosives that contain enough oxygen to sustain the explosion. They are further classified into two subcategories: low-density explosives and high-density explosives. Low-density explosives, such as TNT and RDX, are relatively stable and have a low sensitivity to shock and friction [3]. They are commonly used in military and industrial applications, such as demolition and mining. High-density explosives, such as PETN and HMX, are more energetic than low-density explosives and have a higher sensitivity to shock and friction. They are often used in military applications, such as warheads and shaped charges [4]. High-density explosives can also be used in commercial applications, such as oil well stimulation and seismic exploration. In conclusion, high explosives are a class of substances that undergo a rapid and self-sustaining chemical reaction, resulting in an explosion. They are used for various purposes, including military, industrial, and commercial applications. High explosives can be further classified into primary explosives and secondary explosives, as well as blasting agents and high explosives. Understanding the different types of explosives and their characteristics is crucial for safe handling and use [4].

9.1.1.2 Low Explosives

Low explosives are a class of substances that burn rapidly but do not detonate. They release energy in the form of heat, light, and gas, but the reaction is not self-propagating. This means that the reaction does not generate a shock wave that travels faster than the speed of sound through the material, which is the defining characteristic of high explosives. Low explosives have a controlled burn rate, which makes them useful for specific applications, such as propelling projectiles or creating special effects [5]. Low explosives can be further classified into two categories: pyrotechnics and propellants. Pyrotechnics are substances that burn slowly, producing colorful flames, sparks, and smoke. They are commonly used in fireworks, flares, and signal devices. Pyrotechnics can be further subdivided into two categories: oxidizing agents and reducing agents. Oxidizing agents are substances that provide oxygen to the reaction, while reducing agents are substances that provide fuel to the reaction. Examples of oxidizing agents include potassium perchlorate, barium nitrate, and strontium nitrate, while examples of reducing agents include aluminum, magnesium, and titanium [5].

Propellants, on the other hand, are substances that burn rapidly, producing gas that propels a projectile. They are used in firearms, rockets, and other propulsion systems. Propellants are typically composed of three components: fuel, oxidizer, and binder. The fuel provides the energy to the reaction, the oxidizer provides the oxygen to the reaction, and the binder holds the mixture together. The most common propellants are black powder, smokeless powder, and composite propellants. Black powder, also known as gunpowder, is a low explosive that has been used for centuries as a propellant. It is composed of potassium nitrate, charcoal, and sulfur, and it burns quickly, producing a large amount of gas that propels a projectile. Smokeless powder is a more modern propellant that is composed of nitrocellulose, nitroglycerin, and other additives. It burns more cleanly than black powder and produces less smoke [5]. Composite propellants are a mixture of a fuel, an oxidizer, and a binder, which are combined to form a solid mass. They are commonly used in rockets and missiles. Low explosives are a class of substances that burn rapidly but do not detonate. They are useful for specific applications, such as pyrotechnics and propellants. Pyrotechnics produce colorful flames, sparks, and smoke, while propellants produce gas that propels a projectile. Understanding the different types of low explosives and their characteristics is crucial for safe handling and use.

There are several methods for detecting explosives, and these can be broadly divided into two categories: trace detection and bulk detection. Trace detection methods involve detecting minute amounts of explosive material that may be present on surfaces or in the air. These methods include:

1. **Ion Mobility Spectrometry (IMS)**—It is a method used to separate and identify gas-phase ions based on their mobility in an electric field. It is commonly used for the detection of trace amounts of substances, such as explosives, narcotics, and chemical warfare agents, in air samples. In IMS, ions are generated by ionization of the sample molecules using a radioactive source or a corona discharge [6]. The ions are then introduced into a drift tube, which is filled with a buffer gas, such as nitrogen or helium. As the ions move through the drift tube, they collide with the buffer gas molecules, causing them to slow down and change direction. The rate of this motion is dependent on the size, shape, and charge of the ion, which affects the time it takes for the ion to travel through the drift tube. At the end of the drift tube, the ions are detected by a detector, such as a Faraday

plate or an electron multiplier [6]. The detector measures the ion current and the arrival time of the ions, which are used to calculate the ion mobility. The ion mobility is related to the size, shape, and charge of the ion, which can be used to identify the chemical composition of the sample. The IMS spectrum is a plot of the ion current as a function of the ion mobility, which provides a unique "fingerprint" of the sample. In conclusion, Ion Mobility Spectrometry (IMS) is a technique used to separate and identify gas-phase ions based on their mobility in an electric field. It is commonly used for the detection of trace amounts of substances in air samples. The IMS spectrum provides a unique "fingerprint" of the sample, which can be used to identify its chemical composition [6].

2. **Mass Spectrometry (MS)**—It is an analytical technique used to identify and quantify molecules based on their mass-to-charge ratio (m/z). It is commonly used in chemistry, biochemistry, and other fields to analyze the chemical and physical properties of substances, such as proteins, drugs, and environmental pollutants. In MS, a sample is ionized, which means that it is converted into charged particles, or ions. The ionization can be achieved by various methods, such as electron ionization, electrospray ionization, or matrix-assisted laser desorption/ionization. The resulting ions are then separated based on their m/z ratio using a mass analyzer, which can be a magnetic field, an electric field, or a combination of both the resulting data is called a mass spectrum, which is a plot of the number of ions as a function of their m/z ratio. The mass spectrum provides information about the mass and abundance of the ions in the sample, which can be used to identify the chemical composition of the sample. There are several types of mass analyzers used in MS, including time-of-flight (TOF), quadrupole, ion trap, and Fourier transform ion cyclotron resonance (FT-ICR). Each type of mass analyzer has its own advantages and limitations, depending on the application. In conclusion, it is commonly used in chemistry, biochemistry, and other fields to analyze the chemical and physical properties of substances. The resulting data is called a mass spectrum, which provides information about the mass and abundance of the ions in the sample, which can be used to identify the chemical composition of the sample.
3. **X-ray Diffraction (XRD)**—X-ray Diffraction (XRD) is a powerful analytical technique used to determine the atomic and molecular structure of materials. It works by analyzing the scattering of X-rays by the atoms in a sample, which produces a diffraction pattern that can be used to identify the crystal structure and composition of the material. XRD is commonly used in materials science, chemistry, and physics to study a wide range of materials, including metals, ceramics, minerals, polymers, and biological materials. It is particularly useful for studying crystalline materials, which have a regular, repeating arrangement of atoms that produces a distinctive diffraction pattern. The XRD technique involves shining a beam of X-rays onto a sample and measuring the angles at which the scattered X-rays are detected. The resulting diffraction pattern is then analyzed to determine the crystal structure and composition of the material. This can be done using a variety of methods, including Bragg's equation, which relates the angles of diffraction to the spacing of the atoms in the crystal lattice.

XRD can also be used to study the properties of materials under different conditions, such as temperature, pressure, or in the presence of other materials [7]. This allows researchers to investigate the behavior of materials in real-world applications and to design new materials with specific properties. Overall, X-ray Diffraction is a powerful tool for understanding the structure and properties of materials at the atomic and molecular level. Its versatility and accuracy make it an essential technique in many fields of scientific research. Bulk detection methods involve detecting larger quantities of explosive material that may be present in luggage, cargo, or vehicles. These methods include:

1. ***X-ray imaging***—X-ray imaging is a widely used medical diagnostic tool that allows doctors to see inside the body without the need for invasive procedures. During an X-ray imaging procedure, the patient is positioned between an X-ray source and a detector. The X-ray source emits a beam of radiation that passes through the body and is detected on the other side. The detector then converts the X-ray signal into a digital image that can be viewed on a computer screen. X-ray imaging is particularly useful for visualizing bones and other dense tissues, because they absorb more X-rays than soft tissues. This makes it an effective tool for diagnosing fractures, dislocations, and other bone-related injuries and conditions. X-ray imaging is generally considered safe when used properly, although there is a small risk of exposure to radiation. To minimize this risk, doctors only use X-ray imaging when necessary and take steps to limit the amount of radiation exposure. Overall, X-ray imaging is a valuable tool for diagnosing a wide range of medical conditions and injuries. Its non-invasive nature and ability to produce detailed images quickly make it an essential tool in modern medicine.
2. ***Neutron activation analysis (NAA)***—It is an analytical technique used to determine the elemental composition of materials. It works by exposing a sample to a neutron source, which causes some of the atoms in the sample to become radioactive. The resulting radioactive isotopes can then be measured to determine the concentration of elements in the sample. NAA is particularly useful for analyzing trace elements, which are present in very small quantities and can be difficult to detect using other methods. It is commonly used in fields such as environmental science, archaeology, and materials science to study a wide range of samples, including soil, water, rocks, artifacts, and biological tissues. The NAA technique involves exposing a sample to a neutron source, such as a nuclear reactor or a neutron generator. The neutrons interact with the atoms in the sample, causing some of them to become radioactive. The resulting radioactive isotopes emit gamma rays, which can be detected and measured using specialized equipment. The gamma rays emitted by the sample are characteristic of the elements present in the sample, allowing researchers to identify and quantify the elements using a process called gamma spectroscopy [8]. This involves measuring the energy and intensity of the gamma rays emitted by the sample and comparing them to known standards. NAA is a powerful technique for analyzing samples with complex matrices, because it is non-destructive and does not require extensive sample

preparation. It is also highly sensitive and can detect trace elements at concentrations as low as parts per billion. Overall, neutron activation analysis is a valuable tool for studying the elemental composition of materials. Its sensitivity and versatility make it an essential technique in many fields of scientific research.

3. ***Infrared spectroscopy***—Infrared spectroscopy is an analytical technique used to study the vibrational modes of molecules. It works by measuring the absorption or transmission of infrared radiation by a sample, which can provide information about the functional groups and chemical bonding present in the molecule. Infrared spectroscopy is commonly used in chemistry, biochemistry, and materials science to study a wide range of materials, including organic and inorganic compounds, polymers, and biological tissues. It is particularly useful for identifying and characterizing the functional groups present in organic compounds, such as alcohols, carboxylic acids, and amines. The infrared spectroscopy technique involves shining an infrared beam through a sample and measuring the amount of radiation absorbed or transmitted by the sample. The resulting spectrum is then analyzed to identify the functional groups and chemical bonding present in the molecule [9]. The absorption or transmission of infrared radiation by a molecule depends on the frequency of the radiation and the vibrational modes of the molecule. Different functional groups and chemical bonding arrangements have characteristic vibrational frequencies, which can be identified by their unique absorption or transmission patterns in the infrared spectrum. Infrared spectroscopy can also be used to study the properties of materials under different conditions, such as temperature or pressure. This allows researchers to investigate the behavior of materials in real-world applications and to design new materials with specific properties. Overall, infrared spectroscopy is a powerful tool for studying the structure and properties of molecules and materials.

Metal-organic frameworks (MOFs) are a promising class of materials that have gained a lot of attention in recent years due to their unique properties. These materials are made up of metal ions or clusters that are linked together by organic ligands to form a three-dimensional porous structure. The resulting material has a high surface area, tunable pore size, and high selectivity for certain molecules. One potential application of MOFs is in the detection of explosives. Explosives are a major threat to national security, and there is a need for more sensitive and selective detection methods [10–13].

MOFs offer several advantages over traditional methods of explosive detection, such as their high surface area and tunable pore size, which can be optimized for specific explosive molecules. Several research groups have explored the use of MOFs for explosive detection. For example, one study used a MOF called UiO-66 to detect trace amounts of TNT vapor. The researchers found that UiO-66 had a high sensitivity for TNT, with a detection limit of 1.4 parts per billion. Another study used a MOF called MOF-74 to detect nitroaromatic compounds, which are commonly found in explosives. The researchers found that MOF-74 had a high selectivity for these compounds and could detect them at concentrations as low as 0.2 parts per million. MOFs can also be used in combination with other detection methods, such as fluorescence.

In one study, researchers functionalized a MOF with a fluorescent molecule that was sensitive to TNT. When TNT was present, the fluorescence of the MOF changed, allowing for easy detection. In addition to their high sensitivity and selectivity, MOFs also offer other advantages over traditional detection methods. For example, they can be easily integrated into portable devices, which could be used in the field for rapid detection of explosives. MOFs are also reusable, which makes them more cost-effective than other detection methods. Despite these advantages, there are still some challenges that need to be addressed before MOFs can be widely used for explosive detection. For example, MOFs can be sensitive to moisture and other environmental factors, which can affect their performance. Researchers are working to develop MOFs that are more stable and robust and can withstand harsh environments.

Metal-organic frameworks (MOFs) have shown great potential in the detection of explosives due to their unique chemical and physical properties. MOFs are highly porous materials composed of metal ions or clusters linked by organic ligands, which form a three-dimensional network of cages and channels. The high surface area, tunable pore size, and chemical stability of MOFs make them attractive for the detection of explosives [14–18] (Fig. 9.1).

The scope of MOF detection of explosives is vast and varied. MOFs have been used to detect a wide range of explosives, including nitroaromatics, nitramines, and peroxides. The detection mechanism of MOFs is based on the interaction between the explosive molecule and the MOF framework [19, 20]. This interaction can be either physical (e.g., adsorption) or chemical (e.g., coordination), depending on the nature of the explosive and the MOF. One of the main advantages of MOFs in the detection of explosives is their high selectivity. MOFs can be designed to selectively detect a specific explosive molecule by tuning the size, shape, and chemical functionality of the pores.

This selectivity is crucial in real-world applications, where the presence of other interfering molecules can complicate the detection process. Another advantage of MOFs is their high sensitivity. MOFs can detect explosives at extremely low concentrations, typically in the parts-per-billion or even parts-per-trillion range. This high sensitivity is due to the large surface area and high affinity of MOFs for explosive molecules. MOFs can also be integrated into various sensing platforms, such as optical, electrical, and mechanical sensors, to provide real-time and remote detection of explosives. These sensing platforms can be tailored to specific applications, such as homeland security, environmental monitoring, and medical diagnostics.

One of the biggest challenges is the stability of MOFs under harsh conditions, such as high temperature, humidity, and acidic or basic environments. Researchers are working to improve the stability of MOFs by designing more robust frameworks and developing protective coatings. The scope of MOF detection of explosives is vast and promising. MOFs have shown high selectivity, sensitivity, and versatility in the detection of various explosives. With further research and development, MOFs

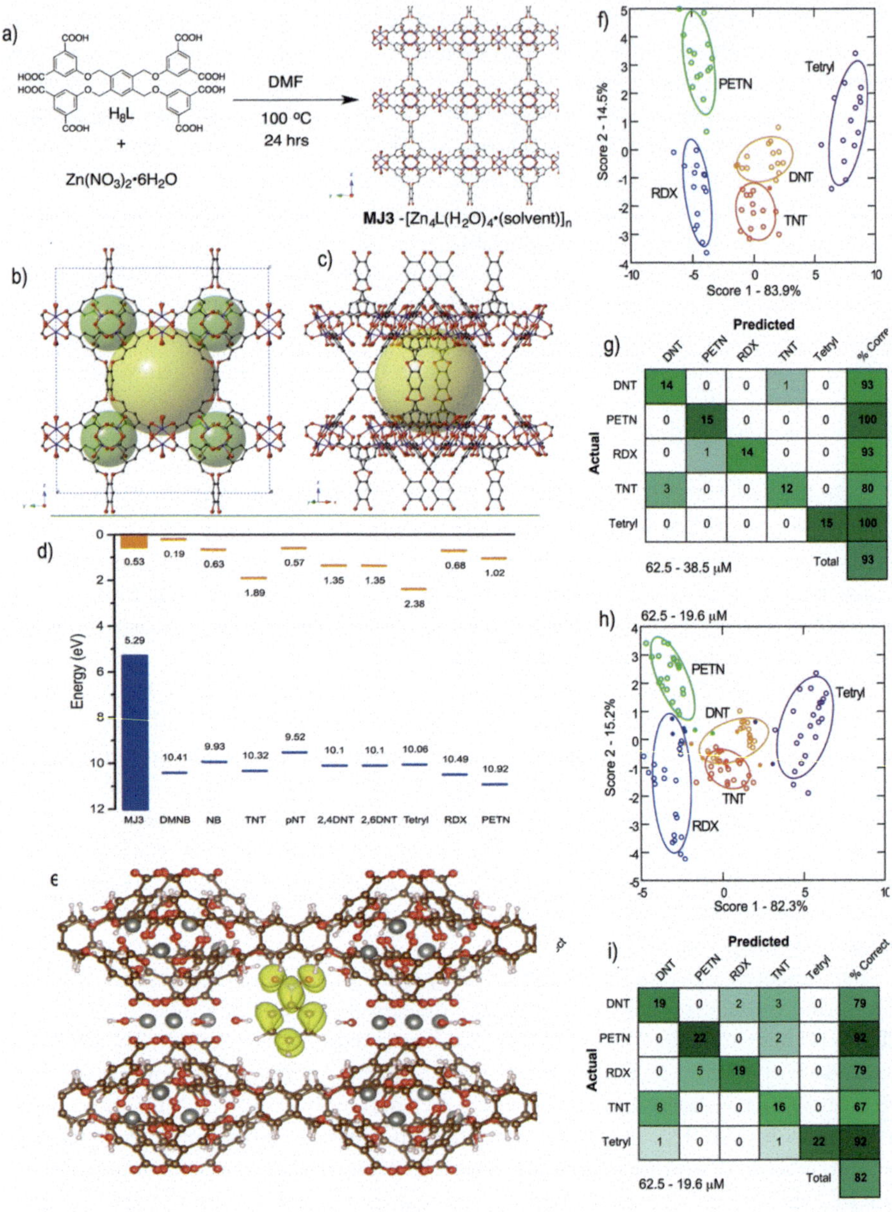

Fig. 9.1 Sensing and discrimination of explosives with a large-pore MOF as part of luminescent array. (Reprint with permission [7] Copyright @2019 ACS)

have the potential to revolutionize the field of explosive sensing and enhance security, safety, and environmental protection.

9.1.2 MOF Synthesis

MOF synthesis is the process of creating new MOFs with specific properties to enhance their performance in various applications. The synthesis of MOFs involves the coordination of metal ions or clusters with organic ligands in a solvent or solvent mixture. The choice of metal ions, ligands, solvents, and reaction conditions can significantly impact the properties of the resulting MOFs, such as their morphology, crystallinity, porosity, and functionality [18–22].

MOF synthesis typically involves several steps, including ligand design, metal coordination, and crystallization. Ligand design is a crucial step in MOF synthesis as it determines the functionality and selectivity of the resulting MOFs. Ligands can be designed to contain various functional groups, such as carboxylic acid, amine, and pyridine, that can interact with metal ions to form coordination bonds. The choice of ligand can also impact the pore size and shape of the MOFs, which can affect their gas adsorption and separation properties. Metal coordination involves the reaction of metal ions or clusters with ligands in a solvent or solvent mixture (Fig. 9.2).

The reaction conditions, such as temperature, pH, and concentration, can significantly impact the coordination between metal ions and ligands, leading to the formation of different MOFs. The choice of metal ions can also impact the properties of the resulting MOFs, such as their thermal stability, redox activity, and magnetic properties.

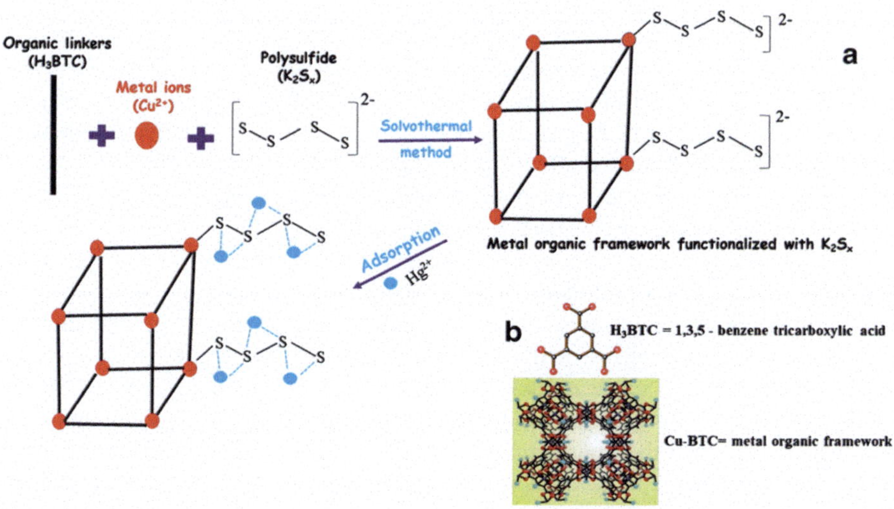

Fig. 9.2 Metal organic frame functionality. (Reprint with permission [23] Copyright @2022 RSC)

Crystallization is the final step in MOF synthesis, where the coordination complexes formed in the previous step are transformed into a crystalline MOF. Crystallization can be achieved through various methods, such as solvothermal, hydrothermal, microwave, or mechanochemical synthesis. The choice of crystallization method can impact the morphology, crystallinity, and porosity of the resulting MOFs [24]. MOF synthesis is a rapidly evolving field with significant potential for developing new and improved MOFs for various applications. Recent advancements in ligand design, metal coordination, and crystallization have enabled the synthesis of MOFs with enhanced properties, such as higher surface area, improved stability, and enhanced selectivity; the use of novel synthesis methods, such as continuous flow synthesis and electrochemical synthesis, has opened up new avenues for MOF synthesis and functionalization. MOF synthesis is a complex and interdisciplinary field that involves the design and synthesis of metal-organic frameworks with specific properties for various applications. The ability to tailor the properties of MOFs through ligand design, metal coordination, and crystallization has significant potential for advancing the field of materials science and developing new technologies for various industries [24].

9.1.3 MOF-Based Sensors

MOF-based sensors are a rapidly emerging class of sensors that are being studied extensively by researchers all over the world. MOFs, or metal-organic frameworks, have a high surface area, which makes them ideal for use in gas sensors, chemical sensors, and other applications. One of the key advantages of MOFs over other materials is their ability to selectively absorb certain gases or chemicals. This makes them highly effective in detecting specific substances, such as carbon dioxide, methane, or even explosives [16–28].

MOFs can be easily modified to target different gases or chemicals, which makes them very versatile. MOF-based sensors work by detecting changes in the electrical conductivity of the material because of the absorption of a gas or chemical. When a gas or chemical is absorbed by the MOF, it causes a change in the electron density of the material, which in turn affects its conductivity. This change can be measured using a variety of techniques, such as electrical impedance spectroscopy or field-effect transistor measurements (Fig. 9.3).

There are many potential applications for MOF-based sensors. For example, they could be used to detect pollutants in the air or water, or to monitor the quality of food or pharmaceutical products. They could also be used in security applications, such as detecting explosives or illicit drugs. One of the challenges of developing MOF-based sensors is ensuring that they are stable and reliable over time. MOFs can be sensitive to changes in temperature, humidity, and other environmental factors, which can affect their performance [30–34]. Researchers are working to overcome these challenges by developing new MOF structures that are more stable and robust, as well as by exploring new fabrication techniques.

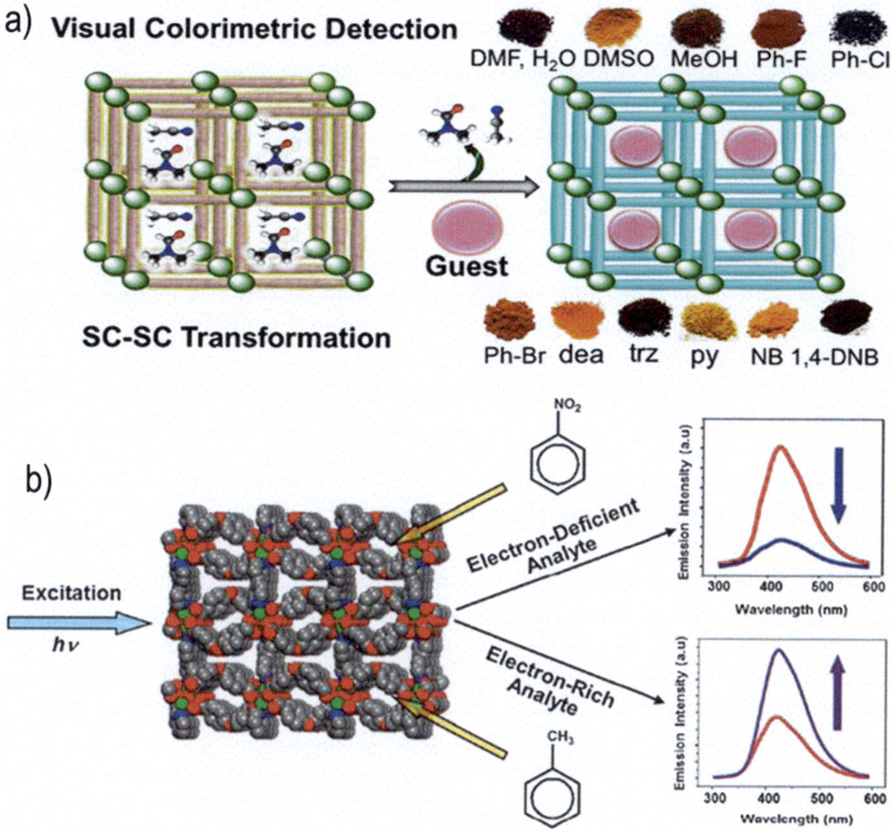

Fig. 9.3 MOF sensor for nitrobenzene (explosive material) detection. (Reprint with permission [29] Copyright @2013 RSC)

9.1.4 MOFs for Trace Detection

MOFs have been extensively studied for their potential to detect trace amounts of toxic gasses, explosives, and drugs, owing to their high surface area and tunable properties. The pore size and chemical functionality of MOFs can be customized to selectively capture and absorb target molecules from complex mixtures, resulting in a highly specific and sensitive method of trace detection. In recent years, researchers have developed several MOFs specifically for the detection of toxic gases, such as nerve agents, mustard gas, and cyanide.

One such MOF-based sensor was able to detect nerve agents like sarin, soman, and tabun at concentrations as low as 10 parts per billion (ppb), exhibiting remarkable selectivity and sensitivity. Similarly, MOFs have been developed for the detection of explosives like TNT and RDX. A MOF-based sensor was able to detect TNT at concentrations as low as 0.5 ppb, and could distinguish TNT from other explosive compounds with remarkable selectivity [34]. In addition to toxic gasses and explosives, MOFs have also shown potential for the detection of drugs. A MOF-based sensor was developed for the detection of cocaine, which was able to detect cocaine at concentrations as low as 10 ppb with high selectivity.

Despite the promising results, there are still challenges to overcome before MOF-based sensors can be practically implemented for trace detection. For instance, MOFs are susceptible to degradation under different conditions like temperature and humidity and may exhibit poor reproducibility that can impact sensor performance.

MOFs present a highly promising avenue for trace detection of toxic gases, explosives, and drugs. Their tunable properties enable selective capture and adsorption of target molecules, creating a specific and sensitive method of trace detection. Nevertheless, challenges such as stability and reproducibility must be overcome before the practical implementation of MOF-based sensors for trace detection can be realized.

9.1.5 MOFs for Real Time Monitoring

MOFs are their high surface area, which can be as high as several thousand square meters per gram, making them ideal for gas storage and separation applications. In addition to their gas storage and separation properties, MOFs are also being investigated for use in real-time monitoring applications. Specifically, researchers are exploring the use of MOFs as sensors for detecting gases, liquids, and even biological molecules [8]. The high surface area of MOFs allows them to interact with a wide range of molecules, making them highly sensitive and selective sensors. One of the most promising applications of MOFs as sensors is in environmental monitoring. For example, MOFs can be used to detect and quantify pollutants in air and water. In air monitoring applications, MOFs can be used to detect harmful gasses such as carbon monoxide, nitrogen oxides, and sulfur dioxide. In water monitoring applications, MOFs can be used to detect heavy metals, organic pollutants, and even bacteria and viruses. Another area where MOFs are being explored for use as sensors is in the medical field.

MOFs can be used to detect and quantify biomolecules such as proteins and nucleic acids, which are important in the diagnosis and treatment of diseases. For example, MOFs can be used to detect cancer biomarkers in blood samples, which can help in the early detection and treatment of cancer. One of the advantages of MOFs as sensors is their ability to provide real-time monitoring. MOFs can be designed to change their properties in response to the presence of a specific molecule, which can be detected using various techniques such as fluorescence, electrical conductivity, or mass spectrometry. This allows for continuous monitoring of the target molecule, providing real-time feedback on the concentration or presence of the molecule.

Researchers are working to improve the stability of MOFs by developing new synthesis methods and modifying their structure. In conclusion, MOFs are a promising class of materials for real-time monitoring applications. Their high surface area, sensitivity, and selectivity make them ideal for detecting gases, liquids, and biomolecules. With further research and development, MOFs have the potential to revolutionize environmental monitoring, medical diagnostics, and other fields where real-time monitoring is critical [8, 14, 35].

9.1.6 MOFs for Selective Adsorption

One of the most promising applications of MOFs is their use in selective adsorption, a process that involves the separation of one component from a mixture based on differences in their affinities for a particular adsorbent material. MOFs are highly porous materials that consist of metal ions or clusters linked together by organic ligands to form a three-dimensional, crystalline structure. The high surface area and tunable properties of MOFs make them ideal for selective adsorption applications, where they can be used to selectively remove specific molecules from a mixture. One of the most well-known applications of MOFs in selective adsorption is their use in gas storage and separation. MOFs can be designed to selectively adsorb certain gases based on their size, shape, and chemical properties.

This has important implications for the storage and transportation of natural gas and other gasses, as it allows for the separation of different components of the gas mixture. Another area where MOFs are being explored for use in selective adsorption is in water treatment. MOFs can be used to selectively remove contaminants from water, such as heavy metals and organic pollutants. This is particularly important in regions where access to clean water is limited, as MOFs can provide an efficient and cost-effective method for water treatment. MOFs are also being investigated for use in catalysis. In catalytic applications, MOFs can be used to selectively adsorb reactant molecules and bring them into proximity with catalytic sites, leading to increased reaction rates and selectivity.

This has important implications for the development of new chemical processes and the production of fine chemicals. One of the advantages of MOFs in selective adsorption applications is their ability to be tailored for specific applications [36]. The properties of MOFs can be modified by changing the metal ions and organic ligands used in their synthesis, allowing for fine-tuning of their selectivity and affinity for specific molecules.

Despite their potential, there are still challenges that need to be addressed before MOFs can be used in real-world selective adsorption applications. One of the major challenges is the scalability of MOF synthesis. MOFs are typically synthesized in small batches in the laboratory, and scaling up the synthesis process to industrial levels is a significant challenge. Researchers are working to develop new synthesis methods that can be scaled up to meet the demands of industrial applications.

9.2 Advancement of MOF in Detection of Explosives

The detection of explosives is a critical issue in the fields of security and defense. Conventional methods for detecting explosives, such as X-ray scanning, sniffer dogs, and chemical analysis, have limitations such as high cost, low sensitivity, and the need for trained personnel. Therefore, there is a need for the development of alternative methods that are cost-effective, highly sensitive, and easy to use.

MOFs are a type of porous material made up of metal ions or clusters linked by organic ligands. The high surface area and tunable pore sizes of MOFs allow for the selective adsorption of target molecules. Additionally, the chemical functionality of MOFs can be tailored to enhance their sensitivity and selectivity towards specific analytes. These properties make MOFs ideal candidates for the detection of explosives. UiO-66 is one of the most widely studied MOFs for explosive detection. This MOF consists of zirconium ions linked by carboxylic acid ligands. The high surface area and large pore size of UiO-66 make it an excellent candidate for the detection of explosives. Researchers have functionalized UiO-66 with various chemical groups to enhance its selectivity towards explosives [37].

These MOFs have shown high sensitivity and selectivity towards a range of explosives. Researchers have also developed MOF-based sensors that can detect explosives in the gas phase or in solution. One of the challenges in the use of MOFs for explosive detection is their stability. MOFs can undergo structural changes in the presence of analytes, which can affect their sensing properties. To overcome this challenge, researchers have developed MOF-based sensors that incorporate signal amplification strategies. For example, researchers have developed MOF-based fluorescence sensors that can detect explosives with high sensitivity and selectivity. These sensors incorporate a fluorescence signal amplification strategy, which enhances the sensitivity of the sensor towards explosives. MOFs have several advantages over other materials for explosive detection.

Firstly, they have a high surface area and tunable pore sizes that allow for the selective adsorption of target molecules. Secondly, their chemical functionality can be tailored to enhance their sensitivity and selectivity towards specific analytes. Thirdly, MOFs can be easily synthesized and scaled up for commercial applications [38]. Lastly, they can be integrated into various sensing platforms such as gas sensors, fluorescence sensors, and electrochemical sensors.

MOFs have shown great potential in the detection of explosives due to their unique properties. Researchers have functionalized MOFs with various chemical groups to enhance their sensitivity and selectivity towards explosives. MOF-based sensors have been developed that can detect explosives in the gas phase or in solution. The development of MOF-based sensors for explosive detection has the potential to revolutionize the field of security and defense by providing cost-effective, highly sensitive, and easy-to-use methods for detecting explosives.

9.3 Discussion

Metal-organic frameworks (MOFs) are a class of porous materials that have gained widespread attention for various applications, including sensing and detection. One such application is the detection of explosives, which has become increasingly important for security and safety purposes. MOFs offer several advantages for explosive detection, including high surface area, tunable pore sizes, and the ability to selectively capture specific molecules. In recent years, several studies have demonstrated the potential of MOFs for explosive detection. For example, a study conducted by He et al. (2019) developed a MOF-based sensor that could detect various explosives, including nitroaromatics and nitramines, with high sensitivity and selectivity [39]. Similarly, a study by Ahamad et al. (2019) developed a MOF-based sensor that could selectively detect TNT, a well-known explosive [39, 40]. One advantage of MOF-based sensors for explosive detection is their ability to selectively capture specific molecules. MOFs can be tailored to selectively capture specific explosives, reducing the risk of false positives or false negatives. Additionally, MOFs can be integrated into portable devices, making them suitable for field applications. For example, MOFs can be sensitive to humidity and may degrade over time, which could affect their performance. Additionally, the synthesis of MOFs can be complex and time-consuming, which could limit their scalability and commercial viability. MOFs hold great promise for the detection of explosives. With further research and development, MOF-based sensors could provide a faster, more reliable, and cost-effective alternative to traditional methods of explosive detection, ultimately helping to improve security and safety.

9.4 Conclusion

In conclusion, the unique properties of MOFs have positioned them as a promising class of materials for the detection of explosives. The high surface area of MOFs provides ample space for adsorption of target molecules, while the tunable pore size and selective molecule adsorption allow for high sensitivity and selectivity [41]. The use of MOFs as explosive sensors offers significant advantages over traditional detection methods, such as their ability to detect trace amounts of explosives with high accuracy and reproducibility. Despite the considerable progress made in MOF design and synthesis, challenges such as stability and reproducibility must be addressed for their successful implementation in real-world scenarios. Recent research has focused on the development of functionalized MOFs, novel sensing platforms, and advanced characterization techniques to overcome these challenges.

Advancements in MOF-based explosive detection have significant implications for public safety, environmental monitoring, and military operations. The ability to detect explosives with high sensitivity, selectivity, and reproducibility can significantly enhance threat detection and response capabilities. Further research is required to improve the stability, selectivity, and reproducibility of MOFs under varying environmental conditions to fully realize their potential in explosive detection.

References

1. Kreno LE, Leong K, Farha OK, Allendorf M, Van Duyne RP, Hupp JT. Metal–organic framework materials as chemical sensors. Chem Rev. 2012;112(2):1105–25.
2. Lu W, Wei Z, Gu ZY, Liu TF, Park J, Park J, et al. Tuning the structure and function of metal–organic frameworks via linker design. Chem Soc Rev. 2014;43(16):5561–93.
3. Li JR, Kuppler RJ, Zhou HC. Selective gas adsorption and separation in metal–organic frameworks. Chem Soc Rev. 2009;38(5):1477–504.
4. Horcajada P, Gref R, Baati T, Allan PK, Maurin G, Couvreur P, et al. Metal–organic frameworks in biomedicine. Chem Rev. 2012;112(2):1232–68.
5. Lee J, Farha OK, Roberts J, Scheidt KA, Nguyen ST, Hupp JT. Metal–organic framework materials as catalysts. Chem Soc Rev. 2009;38(5):1450–9.
6. Xie Z, Ma M, Fan Y. Metal–organic frameworks for analytical chemistry: from sample collection to chromatographic separation. Acc Chem Res. 2016;49(10):2221–30.
7. Jurcic M, Peveler WJ, Savory CN, Bučar DK, Kenyon AJ, Scanlon DO, Parkin IP. Sensing and discrimination of explosives at variable concentrations with a large-pore MOF as part of a luminescent array. ACS Appl Mater Interfaces. 2019;11(12):11618–26.
8. Hu Z, Deibert BJ, Li J. Luminescent metal–organic frameworks for chemical sensing and explosive detection. Chem Soc Rev. 2014;43(16):5815–40.
9. Nagarkar SS, Joarder B, Chaudhari AK, Mukherjee S, Ghosh SK. Highly selective detection of nitro explosives by a luminescent metal–organic framework. Angew Chem. 2013;125(10):2953–7.
10. Cheng T, Hu J, Zhou C, Wang Y, Zhang M. Luminescent metal-organic frameworks for nitro explosives detection. SCIENCE CHINA Chem. 2016;59:929–47.
11. Tian D, Li Y, Chen RY, Chang Z, Wang GY, Bu XH. A luminescent metal–organic framework demonstrating ideal detection ability for nitroaromatic explosives. J Mater Chem A. 2014;2(5):1465–70.
12. Wang B, Lv XL, Feng D, Xie LH, Zhang J, Li M, et al. Highly stable Zr (IV)-based metal–organic frameworks for the detection and removal of antibiotics and organic explosives in water. J Am Chem Soc. 2016;138(19):6204–16.
13. Li WJ, Lü J, Gao SY, Li QH, Cao R. Electrochemical preparation of metal–organic framework films for fast detection of nitro explosives. J Mater Chem A. 2014;2(45):19473–8.
14. Asha KS, Bhattacharyya K, Mandal S. Discriminative detection of nitro aromatic explosives by a luminescent metal–organic framework. J Mater Chem C. 2014;2(47):10073–81.
15. Gole B, Bar AK, Mukherjee PS. Fluorescent metal–organic framework for selective sensing of nitroaromatic explosives. Chem Commun. 2011;47(44):12137–9.
16. Wang S, Wang Q, Feng X, Wang B, Yang L. Explosives in the cage: metal–organic frameworks for high-energy materials sensing and desensitization. Adv Mater. 2017;29(36):1701898.
17. Yao RX, Cui X, Jia XX, Zhang FQ, Zhang XM. A luminescent zinc (II) metal–organic framework (MOF) with conjugated π-electron ligand for high iodine capture and nitro-explosive detection. Inorg Chem. 2016;55(18):9270–5.
18. Lan A, Li K, Wu H, Olson DH, Emge TJ, Ki W, et al. A luminescent microporous metal–organic framework for the fast and reversible detection of high explosives. Angew Chem. 2009;121(13):2370–4.
19. Wang ZJ, Qin L, Chen JX, Zheng HG. H-bonding interactions induced two isostructural cd (II) metal–organic frameworks showing different selective detection of nitroaromatic explosives. Inorg Chem. 2016;55(21):10999–1005.
20. He H, Song Y, Sun F, Zhao N, Zhu G. Sorption properties and nitroaromatic explosives sensing based on two isostructural metal–organic frameworks. Cryst Growth Des. 2015;15(4):2033–8.
21. Qiu ZJ, Fan ST, Xing CY, Song MM, Nie ZJ, Xu L, et al. Facile fabrication of an AIE-active metal–organic framework for sensitive detection of explosives in liquid and solid phases. ACS Appl Mater Interfaces. 2020;12(49):55299–307.

22. Nagarkar SS, Desai AV, Ghosh SK. A fluorescent metal–organic framework for highly selective detection of nitro explosives in the aqueous phase. Chem Commun. 2014;50(64):8915–8.
23. Nozohour Yazdi M, Yamini Y, Asiabi H, Alizadeh A. A metal organic framework prepared from benzene-1, 3, 5-tricarboxylic acid and copper (II), and functionalized with various polysulfides as a sorbent for selective sorption of trace amounts of heavy metal ions. Microchim Acta. 2018;185:1–8.
24. Jurcic M, Peveler WJ, Savory CN, Scanlon DO, Kenyon AJ, Parkin IP. The vapour phase detection of explosive markers and derivatives using two fluorescent metal–organic frameworks. J Mater Chem A. 2015;3(12):6351–9.
25. Yang J, Wang Z, Hu K, Li Y, Feng J, Shi J, Gu J. Rapid and specific aqueous-phase detection of nitroaromatic explosives with inherent porphyrin recognition sites in metal–organic frameworks. ACS Appl Mater Interfaces. 2015;7(22):11956–64.
26. Zeng XS, Xu HL, Xu YC, Li XQ, Nie ZY, Gao SZ, Xiao DR. A series of porous interpenetrating metal–organic frameworks based on fluorescent ligands for nitroaromatic explosive detection. Inorg Chem Front. 2018;5(7):1622–32.
27. Yinon J. Forensic and environmental detection of explosives. Wiley; 1999.
28. Steinfeld JI, Wormhoudt J. Explosives detection: a challenge for physical chemistry. Annu Rev Phys Chem. 1998;49(1):203–32.
29. Ma D, Li B, Zhou X, Zhou Q, Liu K, Zeng G, et al. A dual functional MOF as a luminescent sensor for quantitatively detecting the concentration of nitrobenzene and temperature. Chem Commun. 2013;49(79):8964–6.
30. Yinon J. Field detection and monitoring of explosives. TrAC Trends Anal Chem. 2002;21(4):292–301.
31. Fu HR, Yan LB, Wu NT, Ma LF, Zang SQ. Dual-emission MOF⊃ dye sensor for ratiometric fluorescence recognition of RDX and detection of a broad class of nitro-compounds. J Mater Chem A. 2018;6(19):9183–91.
32. Zhang F, Wang Y, Chu T, Wang Z, Li W, Yang Y. A facile fabrication of electrodeposited luminescent MOF thin films for selective and recyclable sensing of nitroaromatic explosives. Analyst. 2016;141(14):4502–10.
33. Chai HM, Zhang GQ, Jiao CX, Ren YX, Gao LJ. A multifunctional Tb-MOF detector for H2O2, Fe3+, Cr2O72–, and TPA explosive featuring coexistence of binuclear and tetranuclear clusters. ACS Omega. 2020;5(51):33039–46.
34. Lin XM, Niu JL, Lin J, Hu L, Zhang G, Cai YP. A luminescent Tb (III)-MOF based on pyridine-3, 5-dicarboxylic acid for detection of nitroaromatic explosives. Inorg Chem Commun. 2016;72:69–72.
35. Qian JJ, Qiu LG, Wang YM, Yuan YP, Xie AJ, Shen YH. Fabrication of magnetically separable fluorescent terbium-based MOF nanospheres for highly selective trace-level detection of TNT. Dalton Trans. 2014;43(10):3978–83.
36. Singha DK, Majee P, Mondal SK, Mahata P. Visible detection of explosive nitroaromatics facilitated by a large stokes shift of luminescence using europium and terbium doped yttrium-based MOFs. RSC Adv. 2015;5(123):102076–84.
37. Lv R, Wang J, Zhang Y, Li H, Yang L, Liao S, et al. An amino-decorated dual-functional metal–organic framework for highly selective sensing of Cr (iii) and Cr (vi) ions and detection of nitroaromatic explosives. J Mater Chem A. 2016;4(40):15494–500.
38. Kim TK, Lee JH, Moon D, Moon HR. Luminescent Li-based metal–organic framework tailored for the selective detection of explosive nitroaromatic compounds: direct observation of interaction sites. Inorg Chem. 2013;52(2):589–95.
39. He N, Gao M, Shen D, Li H, Han Z, Zhao P. Rapid visual detection of nitroaromatic explosives using a luminescent europium-organic framework material. Forensic Sci Int. 2019;297:1–7.
40. Ahamad MN, Shahid M, Ahmad M, Sama F. Cu (II) MOFs based on bipyridyls: topology, magnetism, and exploring sensing ability toward multiple nitroaromatic explosives. ACS Omega. 2019;4(4):7738–49.
41. Dwivedi B, Mishra PK, Mishra V. Chiral metal-organic frameworks for asymmetrical catalysis. In Metal-organic frameworks (MOFs) as catalysts, Singapore: Springer Nature Singapore, 2022;493–513.

Metal-Organic Frameworks for Detection of Gunshot Residual

10

Abstract

The detection of gunshot residue (GSR) is an essential tool for forensic investigations and criminal justice systems. Traditional methods for GSR detection, such as scanning electron microscopy (SEM), energy-dispersive X-ray spectroscopy (EDX), and atomic absorption spectroscopy (AAS), are highly accurate but are also expensive and time-consuming and require specialized equipment and expertise. Recently, metal-organic frameworks (MOFs) have emerged as a promising alternative for GSR detection due to their high sensitivity, selectivity, and stability. MOFs are composed of metal ions or clusters connected by organic linkers to form a three-dimensional porous network with large surface areas. MOFs are versatile materials that can be engineered to selectively adsorb specific inorganic compounds, making them highly selective for GSR detection. MOFs can also detect very low levels of GSR particles, which is essential for the detection of trace amounts of GSR in complex samples. Additionally, MOFs are stable under a wide range of environmental conditions and can withstand exposure to various chemicals and solvents, making them ideal for use in harsh conditions, such as in forensic investigations. The mechanism of detection of GSR using MOFs is based on the fact that GSR particles contain a mixture of inorganic compounds, such as lead, barium, and antimony, which can be selectively adsorbed by MOFs. When exposed to a sample containing GSR particles, the MOF material selectively absorbs the inorganic compounds present in the GSR particles. This process leads to a measurable change in the dielectric constant of the MOF material, which can be detected by various methods. The performance evaluation of MOFs for GSR detection has shown promising results in terms of sensitivity, selectivity, and reproducibility. MOFs have shown high sensitivity for the detection of GSR particles, which is essential for the detection of trace amounts of GSR in complex samples.

V. Mishra, A. Mishra, *Metal-Organic Frameworks as Forensic Detectors*,
https://doi.org/10.1007/978-981-96-2147-7_10

10.1 Introduction

Firearms are weapons that use gunpowder or other explosives to propel a projectile, such as a bullet, out of a barrel and towards a target. Firearms have been used for centuries for hunting, self-defense, and military purposes, and have evolved over time to become more accurate, powerful, and versatile. There are many different types of firearms, including pistols, rifles, shotguns, and machine guns. Each type of firearm has its own characteristics and uses, depending on the situation and the user's preferences. Pistols are small, easily concealable firearms that are commonly used for self-defense or as a backup weapon. They are typically fired with one hand and have a short range and low accuracy compared to other firearms. Rifles are long-barreled firearms that are designed for accuracy and range. They are commonly used for hunting, target shooting, and military purposes, and can be fitted with a variety of accessories, such as scopes, bipods, and suppressors. Shotguns are firearms that fire a cluster of small pellets, or shot, instead of a single bullet. They are commonly used for hunting birds or small game and are also popular for home defense because of their widespread and stopping power. Machine guns are fully automatic firearms that are designed to fire multiple rounds with a single pull of the trigger. They are primarily used by military and law enforcement personnel and are heavily regulated or illegal for civilian use in many countries. Firearms can be divided into two main categories: handguns and long guns. Handguns are firearms that are designed to be held in one hand, while long guns are designed to be held with both hands and rest against the shoulder for stability. Handguns are typically smaller and more easily concealable than long guns, while long guns are generally more accurate and powerful. Firearms use a variety of mechanisms to fire a projectile, including single-action, double-action, and semi-automatic mechanisms. Single-action firearms require the user to manually cock the hammer before firing, while double-action firearms can be fired by simply pulling the trigger. Semi-automatic firearms use the energy from the firing of one round to automatically load and fire the next round, allowing for rapid fire without the need to manually reload between shots [1]. The development of firearms has had a significant impact on human history, from the use of early firearms in warfare to the modern-day debate over gun control. Firearms have been used to defend against invaders, expand empires, and overthrow governments. They have also been used for criminal purposes, such as robbery and murder, and have been the subject of controversy and regulation in many countries. In recent years, there has been significant debate over the role of firearms in society, with proponents arguing that firearms are essential for self-defense and freedom, and opponents arguing that they contribute to violence, crime, and accidents. Many countries have enacted laws to regulate the possession and use of firearms, including background checks, waiting periods, and restrictions on certain types of firearms. Overall, firearms are powerful and potentially deadly weapons that have played a significant role in human history. While they can be used for both legitimate and criminal purposes, it is important to use them responsibly and to follow all applicable laws and regulations.

10.1.1 Gunshot Residue

Gunshot residue (GSR) is a type of forensic evidence that is left behind after a firearm is discharged. It consists of microscopic particles that are expelled from the firearm's barrel and can be found on the shooter's hands, clothing, and other surfaces in the vicinity of the shooting. GSR is composed of a mixture of materials, including gunpowder, primer residue, and bullet lubricants. When a firearm is fired, these materials are heated to high temperatures and expelled from the barrel, producing a cloud of particles that can be deposited on nearby surfaces. Forensic investigators can use GSR analysis to determine whether a person has recently fired a firearm, which can be useful in criminal investigations. GSR analysis involves collecting samples from the hands, clothing, and other surfaces of a person who is suspected of firing a gun and analyzing them for the presence of gunshot residue. There are several different methods for collecting and analyzing GSR samples, including swabbing the hands and clothing with adhesive tape or conducting a chemical test to detect the presence of gunshot residue. The samples are then examined under a microscope to identify the characteristic particles of GSR. However, the presence of GSR does not necessarily indicate that a person has fired a gun. GSR can also be found on the hands and clothing of people who have been near a shooting, such as witnesses or first responders. In addition, GSR analysis can be affected by a few factors, such as the type of firearm used, the distance from which the gun was fired, and the weather conditions at the time of the shooting. It is important for forensic investigators to take these factors into account when analyzing GSR evidence. Overall, gunshot residue analysis is a valuable tool for forensic investigators in criminal investigations. While it is not always conclusive, it can provide important evidence that can help to identify suspects and build a case against them (Fig. 10.1).

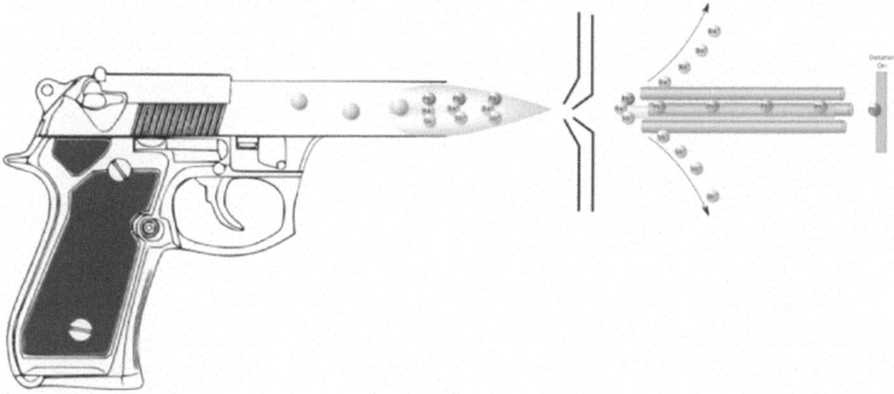

Fig. 10.1 Gunshot residue (GSR) analysis by single particle inductively coupled plasma mass spectrometry (spICP-MS). (Reprint with permission from ref. [2] copyright 2018 Elsevier)

10.1.2 Identification Methods

Gunshot residue (GSR) is a type of forensic evidence that can provide important information about a shooting, including the type of firearm used and the distance from which the gun was fired. There are several methods for identifying GSR, each with its own advantages and limitations.

1. **Scanning Electron Microscopy (SEM)-** Scanning Electron Microscopy (SEM) is a powerful imaging technique used to examine the surface structure and composition of materials at high magnification. It is widely used in materials science, biology, and other fields to study a wide range of materials, from biological tissues to metals and ceramics [2–6]. The SEM technique works by scanning a focused beam of electrons over the surface of the sample being studied. The electrons interact with the atoms in the sample, producing secondary electrons and other signals that can be detected and used to produce an image of the surface. SEM can produce images at very high magnifications, allowing researchers to examine the surface structure of materials at the nanoscale level. It can also be used to analyze the composition of materials, by detecting and measuring the energy of the backscattered electrons or X-rays produced by the sample. One of the key advantages of SEM is its ability to produce high-resolution images of the sample's surface. This is achieved by focusing the electron beam using a series of electromagnetic lenses, which can produce a very small probe size and high current density (Fig. 10.2).

   The resulting images can reveal details such as surface roughness, crystal structure, and defects in the material. In addition to imaging, SEM can be used for a variety of other applications, including elemental analysis, chemical mapping, and electron diffraction. These techniques can provide detailed information about the composition and structure of the material being studied and are useful for a wide range of scientific and industrial applications. SEM is a complex and sophisticated technique that requires specialized equipment and trained operators. The sample being studied must be prepared carefully, to ensure that it

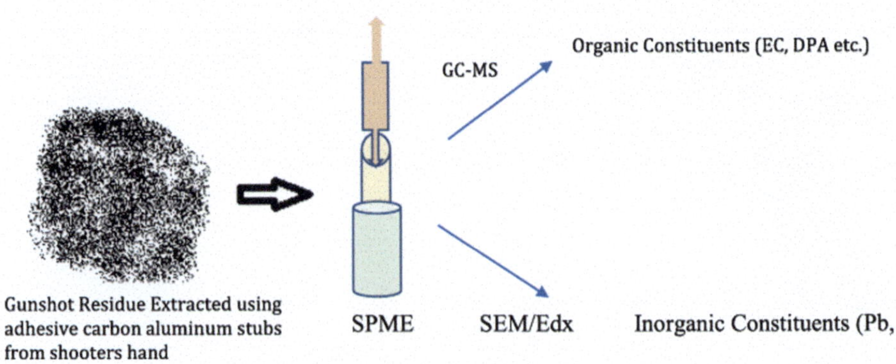

Fig. 10.2 Gunshot residue analysis technique using solid-phase micro extraction. *SPME* Solid phase microextraction, *EC* ethyl centralite, *DPA* diphenyl amine [7]

is free of contamination and properly mounted on the sample holder [7]. The sample must also be conductive or coated with a conductive material, to allow the electron beam to interact with the surface. Overall, Scanning Electron Microscopy is a valuable tool for studying the surface structure and composition of materials at high magnification. Its ability to produce high-resolution images and perform a range of analytical techniques makes it an essential technique in many fields of scientific research and industrial applications.

2. **Atomic Absorption Spectroscopy (AAS)** Atomic Absorption Spectroscopy (AAS) is a technique used to determine the concentration of metallic elements in a sample. It is widely used in analytical chemistry, environmental science, and other fields to measure the levels of metals in a wide range of materials, including soil, water, and biological tissues. The AAS technique works by exposing the sample to a beam of light at a specific wavelength, which causes the atoms in the sample to absorb the light energy. The amount of absorption is proportional to the concentration of the metallic element in the sample [8–10]. AAS can be used to detect a wide range of metallic elements, including lead, copper, zinc, and iron. It is particularly useful for analyzing trace elements, which are present in very small quantities and can be difficult to detect using other methods. There are two main types of AAS: flame AAS and graphite furnace AAS. Flame AAS involves atomizing the sample in a flame and analyzing the absorption of the light by the metallic element.
3. **Inductively Coupled Plasma Mass Spectrometry (ICP-MS)** Inductively Coupled Plasma Mass Spectrometry (ICP-MS) is a powerful analytical technique used to determine the concentration of metallic and non-metallic elements in a sample. It is widely used in environmental science, geochemistry, and other fields to study a wide range of materials, including soil, water, and biological tissues [6]. The ICP-MS technique works by atomizing the sample and ionizing the resulting atoms using an inductively coupled plasma (ICP). The ions are then separated and detected using a mass spectrometer, which can identify the different elements present in the sample and measure their concentrations. MS can detect a wide range of elements, from lithium to uranium, and can measure concentrations as low as parts per trillion. It is particularly useful for analyzing trace elements, which are present in very small quantities and can be difficult to detect using other methods.

There are several advantages to using ICP-MS over other analytical techniques, including its high sensitivity, accuracy, and speed. It can also be used to analyze a wide range of sample types, including solids, liquids, and gases. ICP-MS is a complex and sophisticated technique that requires specialized equipment and trained operators. The sample being studied must be prepared carefully, to ensure that it is free of contamination and properly dissolved or digested prior to analysis. One of the key challenges in using ICP-MS is dealing with potential interferences from other elements in the sample. These interferences can result in false positives or negatives and can affect the accuracy and reliability of the analysis. To address this issue, different methods can be used, such as collision or reaction cells, to separate the ions and reduce the interferences.

Overall, Inductively Coupled Plasma Mass Spectrometry is a valuable tool for studying the composition of materials at the atomic level [11]. Its ability to detect and measure a wide range of elements at low concentrations makes it an essential technique in many fields of scientific research and industrial applications.

4. **Neutron Activation Analysis (NAA)** Neutron Activation Analysis (NAA) is an analytical technique used to determine the concentration of elements in a sample. It works by exposing the sample to a neutron source, which causes some of the atoms in the sample to become radioactive. The resulting radioactive isotopes can then be measured to determine the concentration of elements in the sample. NAA is particularly useful for analyzing trace elements, which are present in very small quantities and can be difficult to detect using other methods. It is commonly used in materials science, environmental science, and other fields to study a wide range of materials, including soil, water, and biological tissues. The NAA technique involves exposing the sample to a neutron source, such as a nuclear reactor or a neutron generator. The neutrons interact with the atoms in the sample, causing some of them to become radioactive [8, 12, 13]. The resulting radioactive isotopes emit gamma rays, which can be detected and measured using specialized equipment.
5. **Fourier Transform Infrared (FTIR) Spectroscopy** Fourier Transform Infrared (FTIR) Spectroscopy is an analytical technique used to study the vibrational modes of molecules. It works by measuring the absorption or transmission of infrared radiation by a sample, which can provide information about the functional groups and chemical bonding present in the molecule. FTIR is commonly used in chemistry, biochemistry, and materials science to study a wide range of materials, including organic and inorganic compounds, polymers, and biological tissues. It is particularly useful for identifying and characterizing the functional groups present in organic compounds, such as alcohols, carboxylic acids, and amines [14]. The FTIR technique involves shining an infrared beam through a sample and measuring the amount of radiation absorbed or transmitted by the sample. The resulting spectrum is then analyzed to identify the functional groups and chemical bonding present in the molecule. The absorption or transmission of infrared radiation by a molecule depends on the frequency of the radiation and the vibrational modes of the molecule. Different functional groups and chemical bonding arrangements have characteristic vibrational frequencies, which can be identified by their unique absorption or transmission patterns in the infrared spectrum. FTIR can also be used to study the properties of materials under different conditions, such as temperature or pressure. This allows researchers to investigate the behavior of materials in real-world applications and to design new materials with specific properties. One of the key advantages of FTIR is its ability to analyze samples non-destructively. This means that the sample can be analyzed without the need for extensive sample preparation, which can save time and reduce the risk of sample contamination. Overall, Fourier Transform Infrared Spectroscopy is a powerful tool for studying the structure and properties of molecules and materials [15]. Its ability to provide detailed information about the

functional groups and chemical bonding in a sample makes it an essential technique in many fields of scientific research.

6. **X-ray Fluorescence (XRF)** X-ray Fluorescence (XRF) is an analytical technique used to determine the elemental composition of a material. It works by analyzing the X-rays emitted by a sample when it is exposed to a beam of X-rays. XRF is commonly used in materials science, geology, and other fields to study a wide range of materials, including metals, ceramics, minerals, and biological tissues. It is particularly useful for analyzing the composition of materials in a non-destructive manner, as it does not require the sample to be destroyed or altered in any way. The XRF technique involves exposing the sample to a beam of X-rays, which causes the atoms in the sample to emit characteristic X-rays [11]. These X-rays are then detected and analyzed to determine the elemental composition of the sample. One of the key advantages of XRF is its ability to analyze a wide range of elements, from magnesium to uranium. It is also relatively fast and easy to use, making it a popular analytical technique in many industries.

10.2 Introduction to Metal-Organic Frameworks (MOFs)

A gunshot is the sound produced by the discharge of a firearm. It is typically a loud, sharp noise, and it can vary in volume depending on the type of firearm and the environment in which it is fired. Gunshots are often associated with violence and crime, and they can cause serious injury or death if the bullet strikes a person or animal. In addition to producing a loud noise, a gunshot also produces a visible muzzle flash and a recoil force that can be felt by the shooter. The muzzle flash is caused by the hot gases and burning gunpowder that are expelled from the barrel of the firearm, while the recoil force is caused by the momentum of the bullet leaving the barrel. Gunshots can be classified based on the type of firearm used, the size and caliber of the bullet, and the number of shots fired. For example, a gunshot from a handgun will typically produce a different sound than a gunshot from a rifle, and a single shot will sound different than a volley of shots fired in rapid succession. Metal-organic frameworks (MOFs) are a class of materials that have received significant attention in recent years due to their tunable, highly porous nature, and potential for various applications [16]. One such promising application is the detection of gunshot residue (GSR) in forensic investigations. GSR is a complex mixture of inorganic and organic compounds that are released when a firearm is discharged. These compounds can be deposited on the shooter's hands, clothing, and nearby surfaces, making them an invaluable tool for identifying and linking suspects to a crime scene. However, traditional GSR analysis methods are often laborious, time-consuming, and costly, involving sophisticated laboratory techniques such as scanning electron microscopy and inductively coupled plasma mass spectrometry.

Gunshot residue (GSR) is a type of physical evidence that is left behind after a firearm is discharged. It consists of small particles that are expelled from the barrel of the gun along with the bullet and the hot gases produced by the combustion of gunpowder. GSR particles are typically composed of lead, barium, and antimony, which are the primary components of the gunpowder used in most firearms. They are very small, typically measuring only a few microns in diameter, and can be difficult to detect without specialized equipment. Forensic investigators can use a variety of methods to collect and analyze GSR, including swabbing the hands and clothing of a suspect, using adhesive tape to collect particles from surfaces, or using electronic sensors to detect the presence of GSR in the air. Once collected, the particles can be analyzed using techniques such as scanning electron microscopy or energy-dispersive X-ray spectroscopy. The presence of GSR on a person or at a crime scene can be used as evidence to link a suspect to a particular firearm or to indicate that a firearm was discharged in the area [17]. However, the presence of GSR alone is not sufficient to prove that a person fired a gun, as the particles can also be transferred through contact with surfaces or objects that have come into contact with GSR (Fig. 10.3).

MOFs offer a promising alternative for GSR detection due to their high surface area and chemical selectivity. They are composed of metal ions or clusters linked by organic ligands to create a porous framework. The large surface area of MOFs allows for the adsorption of small molecules, while the functional groups on the organic ligands can selectively interact with specific compounds. Several MOFs have been investigated for GSR detection, including UiO-66, MIL-101, and ZIF-8. UiO-66, a zirconium-based MOF, has been shown to selectively absorb lead, barium, and antimony, the three most common elements present in GSR. MIL-101, a chromium-based MOF, can selectively adsorb organic compounds such as nitroaromatics and nitramines, which are present in gunpowder. ZIF-8, a zinc-based MOF, can selectively adsorb nitroaromatics and nitramines, as well as heavy metals such as lead and barium.

The detection of GSR using MOFs involves the collection of samples from the crime scene or the suspect's hands or clothing. The samples are then extracted using a solvent, and the resulting solution is exposed to the MOF. The MOF selectively adsorbs the GSR compounds, which can then be analyzed using various techniques such as Fourier transform infrared spectroscopy or gas chromatography-mass spectrometry. The use of MOFs for GSR detection offers several advantages over traditional methods. MOFs are highly sensitive and selective, allowing for the detection of trace amounts of GSR compounds. They are also portable and can be used in the field, reducing the time and cost associated with laboratory analysis. Furthermore, MOFs are reusable and can be regenerated by heating or washing, making them more environmentally friendly than traditional methods. In conclusion, MOFs are a highly promising alternative for the detection of GSR in forensic investigations. Their high surface area and chemical selectivity make them highly sensitive and selective for GSR compounds, while their portability, reusability, and eco-friendliness make them a practical and cost-effective tool for forensic scientists. As research in this area continues, MOFs may become a standard tool for GSR detection, revolutionizing forensic science and aiding in criminal investigations [18–22].

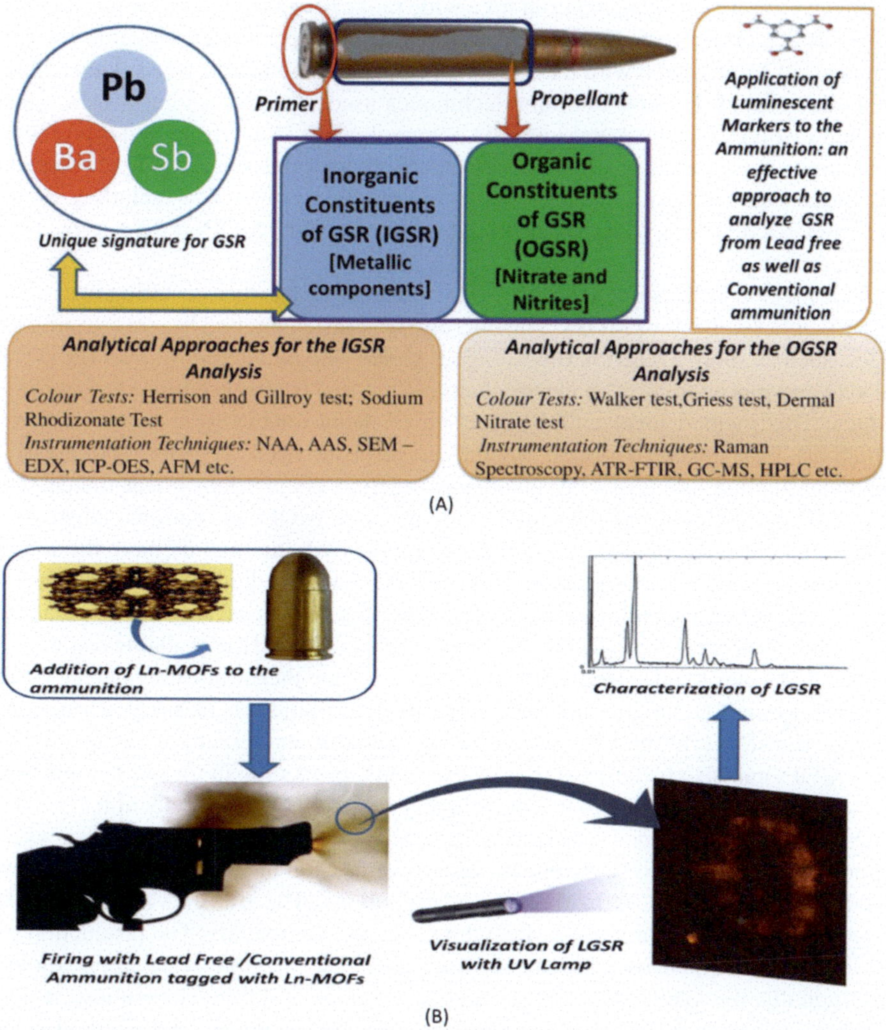

Fig. 10.3 Analytical contributions of lanthanide-based metal-organic frame works as luminescent markers: Recent trends in gunshot residue analysis. (Reprint with permission from ref. [17] copyright 2020 Elsevier)

10.2.1 Synthesis of MOFs for GSR Detection: Methods and Techniques

The synthesis of Metal-Organic Frameworks (MOFs) for the purpose of gunshot residue (GSR) detection is a rapidly developing area of research in forensic science. MOFs are an innovative class of porous materials composed of metal ions or clusters linked by organic ligands. Due to their unique properties, such as high porosity, tunable pore size, and chemical stability, MOFs have gained significant attention in recent years as a promising material for a wide range of sensing applications, including GSR detection. There are numerous methods and techniques available for the synthesis of MOFs, each with its own advantages and limitations. The solvothermal method is one of the most common techniques used for synthesizing MOFs. This method involves heating metal ions and organic ligands in a solvent under pressure to promote the formation of MOF crystals. Solvothermal synthesis is a versatile technique that can be used to synthesize a wide range of MOFs, including those used for GSR detection. Another technique for synthesizing MOFs is the microwave-assisted method.

This method utilizes microwaves to heat the reaction mixture, resulting in a rapid and efficient synthesis of MOFs. The microwave-assisted method has several advantages, such as shorter reaction times, higher yields, and better control over the crystal size and morphology. The electrochemical method is also used for synthesizing MOFs. This method involves the reduction of metal ions to form metal clusters, which are then linked by organic ligands to form the MOF crystals. The electrochemical method offers several advantages, such as better control over the crystal size and morphology and the ability to synthesize MOFs on various substrates. Other methods for synthesizing MOFs include the hydrothermal method, the sonochemical method, and the spray-drying method. Each method has its own advantages and limitations, and the choice of method depends on several factors, such as the desired properties of the MOF, the target application, and the available resources. Once MOFs are synthesized, they are evaluated for their ability to detect GSR [23–25]. Several techniques are used for GSR detection, such as X-ray diffraction, Fourier-transform infrared spectroscopy, and gas chromatography-mass spectrometry. MOFs can also be tailored to selectively adsorb specific inorganic compounds present in GSR particles, making them highly selective for GSR detection.

MOFs are stable under a wide range of environmental conditions and can withstand exposure to various chemicals and solvents, making them ideal for use in harsh conditions, such as in forensic investigations. The use of MOFs for GSR detection also has its limitations, including cost and complexity. The production of MOFs can be expensive, which can limit their use in large-scale applications. Additionally, the synthesis of MOFs can be complex and time-consuming, which can limit their use in practical applications. The sensitivity of MOFs can also be affected by environmental conditions, such as temperature and humidity, which can affect their performance. Finally, the mechanisms and principles of MOFs for GSR detection are still not fully understood, which can limit their optimization for practical applications. In conclusion, the use of MOFs for the detection of GSR offers several advantages over traditional detection methods. MOFs are versatile materials that can be tailored to selectively adsorb specific inorganic compounds and detect very low levels of GSR particles. MOFs are also stable under a wide range of environmental conditions and can withstand exposure to various chemicals and solvents. However, the use of MOFs for GSR detection also has its limitations, including cost and complexity. Further research is needed to fully understand the mechanisms and principles of MOFs for GSR detection and to optimize their sensitivity, selectivity, and stability for practical applications. The potential applications of MOFs in GSR detection are promising and have the potential to revolutionize the field of forensic investigations and criminal justice systems.

These techniques can detect specific chemicals present in GSR, such as lead, barium, and antimony [26]. By detecting these specific chemicals, MOFs can be used to identify the presence of GSR and provide valuable evidence in forensic investigations. In conclusion, the synthesis of MOFs for GSR detection is a rapidly developing area of research in forensic science. The choice of synthesis method depends on several factors, including the desired properties of the MOF and the target application. Once synthesized, MOFs are tested for their ability to detect GSR using various techniques such as X-ray diffraction and gas chromatography-mass spectrometry. With continued research and development, MOFs have the potential to become a powerful tool in GSR detection, helping to solve crimes and bring justice to victims.

10.2.2 Structural Characterization of MOFs: X-Ray Diffraction, NMR Spectroscopy, and Electron Microscopy

Metal-Organic Frameworks (MOFs) have gained significant attention in recent years as promising materials for a wide range of applications, including gas separation, catalysis, and sensing. One of the key factors that determine the properties and performance of MOFs is their structural characteristics. In this article, we will discuss the various techniques used for the structural characterization of MOFs, including X-ray diffraction, NMR spectroscopy, and electron microscopy. X-ray diffraction (XRD) is one of the most commonly used techniques for the structural characterization of MOFs. XRD is a non-destructive technique that allows for the determination of the crystal structure of MOFs. XRD works by directing X-rays through a sample, where they interact with the atoms in the sample and create a diffraction pattern that can be analyzed to determine the crystal structure. XRD can provide information on the unit cell dimensions, the space group, and the orientation of the crystal lattice.

Another technique used for the structural characterization of MOFs is nuclear magnetic resonance (NMR) spectroscopy. NMR spectroscopy is a non-destructive technique that provides information on the local environment of nuclei in a sample. In MOFs, NMR spectroscopy is used to determine the coordination environment of metal ions and the conformation of organic ligands. NMR spectroscopy can also provide information on the dynamics of MOFs, such as the motion of organic ligands within the structure. Electron microscopy is another powerful technique used for the structural characterization of MOFs. Electron microscopy includes several types of microscopies, such as transmission electron microscopy (TEM), scanning electron microscopy (SEM), and high-resolution transmission electron microscopy (HRTEM). These techniques allow for the visualization of MOFs at the nanoscale level [27–29]. Electron microscopy can provide information on the morphology and size of MOF crystals, as well as the distribution of metal ions and organic ligands within the structure. In addition to these techniques, several other techniques are also used for the structural characterization of MOFs, such as infrared spectroscopy (IR), Raman spectroscopy, and thermogravimetric analysis (TGA). Each technique provides unique information on the structure and properties of MOFs, and the choice of technique depends on the specific application.

The information obtained from structural characterization techniques is essential for understanding the properties and performance of MOFs. For example, XRD can provide information on the pore size, surface area, and thermal stability of MOFs, which are important properties for gas separation and catalysis applications. NMR spectroscopy can provide information on the surface chemistry of MOFs, which is important for sensing applications. Electron microscopy can provide information on the morphology and size of MOF crystals, which is important for understanding the growth mechanism of MOFs [30]. Furthermore, structural characterization techniques can also be used to study the properties of MOFs under different conditions, such as varying temperature, pressure, or gas environment.

This information is important for understanding the stability and performance of MOFs in real-world applications. For example, XRD can be used to study the structural changes of MOFs under different gas environments, which is important for gas adsorption and separation applications. In conclusion, the structural characterization of MOFs is essential for understanding their properties and performance in various applications. X-ray diffraction, NMR spectroscopy, and electron microscopy are some of the commonly used techniques for the structural characterization of MOFs. These techniques provide information on the crystal structure, coordination environment, morphology, and size of MOFs. With continued research and development, MOFs have the potential to become a powerful tool in a wide range of applications, including gas separation, catalysis, and sensing.

10.2.3 Physicochemical Characterization of MOFs: Thermal Analysis, Gas Sorption, and Surface Area Measurements

The physicochemical properties of Metal-Organic Frameworks (MOFs) play a critical role in determining their performance for a wide range of applications, including gas separation, catalysis, and sensing. In this article, we will discuss the various techniques used for the physicochemical characterization of MOFs, including thermal analysis, gas sorption, and surface area measurements. Thermal analysis is an important technique used for the physicochemical characterization of MOFs. Thermal analysis includes several techniques, such as thermogravimetric analysis (TGA), differential scanning calorimetry (DSC), and thermal conductivity measurements [32]. TGA is a commonly used technique that measures the weight loss of MOFs as a function of temperature.

This information can be used to determine the thermal stability of MOFs and the presence of any organic or inorganic impurities. DSC is another commonly used technique that measures the heat flow of MOFs as a function of temperature. DSC can provide information on the phase transitions and melting points of MOFs. Thermal conductivity measurements can provide information on the thermal conductivity of MOFs, which is important for understanding their heat transfer properties. Gas sorption is another important technique used for the physicochemical characterization of MOFs. Gas sorption includes several techniques, such as gas adsorption, gas permeation, and gas chromatography. Gas adsorption is a commonly used technique that measures the amount of gas adsorbed onto the surface of MOFs as a function of pressure. Gas adsorption can provide information on the surface area, pore size, and pore volume of MOFs. Gas permeation is another technique that measures the rate of gas diffusion through MOFs. Gas permeation can provide information on the gas selectivity and permeability of MOFs. Gas chromatography is a technique used to separate and analyze the components of a gas mixture. Gas chromatography can be used to study the gas separation properties of MOFs.

Surface area measurements are also important for the physicochemical characterization of MOFs. Surface area measurements include several techniques, such as BET analysis, Langmuir analysis, and mercury intrusion porosimetry [33, 34]. BET analysis is a commonly used technique that measures the surface area of MOFs based on the adsorption of a gas onto the surface of MOFs. Langmuir analysis is another technique that measures the surface area of MOFs based on the adsorption of a gas onto a monolayer on the surface of MOFs. Mercury intrusion porosimetry is a technique that measures the pore size distribution of MOFs based on the intrusion of mercury into the pores of MOFs.

The information obtained from physicochemical characterization techniques is essential for understanding the properties and performance of MOFs. For example, gas sorption measurements can provide information on the surface area, pore size, and pore volume of MOFs, which are important properties for gas separation and catalysis applications. Thermal analysis can provide information on the thermal stability of MOFs, which is important for understanding their performance under high temperature conditions. Surface area measurements can provide information on the accessibility of the pores in MOFs, which is important for understanding their gas adsorption and separation properties. Furthermore, physicochemical characterization techniques can also be used to study the properties of MOFs under different conditions, such as varying temperature, pressure, or gas environment.

This information is important for understanding the stability and performance of MOFs in real-world applications. For example, gas sorption measurements can be used to study the gas adsorption properties of MOFs under different gas environments, which is important for gas separation applications. The physicochemical characterization of MOFs is essential for understanding their properties and performance in various applications. Thermal analysis, gas sorption, and surface area measurements are some of the commonly used techniques for the physicochemical characterization of MOFs [35]. These techniques provide information on the thermal stability, gas adsorption properties, and surface area of MOFs. With continued research and development, MOFs have the potential to become a powerful tool in a wide range of applications, including gas separation, catalysis, and sensing.

10.2.4 Surface Modification of MOFs for Improved GSR Detection: Functionalization, Doping, and Encapsulation

Surface modification of metal-organic frameworks (MOFs) has emerged as a promising strategy to enhance their gas sensing properties. MOFs are highly porous materials composed of metal ions or clusters connected by organic ligands, which result in a large surface area and tunable pore size. However, their sensitivity and selectivity towards gas molecules can be limited by their inherent properties, including weak interactions and poor stability under harsh environmental conditions. Therefore, surface modification techniques have been developed to overcome these limitations and improve the performance of MOF-based gas sensors. Functionalization is one of the most commonly used surface modification techniques for MOFs. It involves the attachment of functional groups, such as –NH2, –COOH, and –OH, onto the surface of MOFs to enhance their interaction with gas molecules.

The functional groups can also provide additional binding sites for target molecules, leading to increased selectivity and sensitivity of the MOF-based sensors. For example, the attachment of –NH2 groups onto the surface of ZIF-8 MOFs has been shown to improve the detection of trace amounts of methyl mercaptan, a common gas associated with halitosis and flatulence. Doping is another surface modification technique that involves the incorporation of foreign atoms or molecules into the MOF structure. This can alter the electronic properties of MOFs and enhance their interaction with gas molecules. For instance, the doping of ZIF-8 MOFs with transition metal ions, such as Co2+, Ni2+, and Cu2+, has been reported to improve their sensitivity towards various gases, including ammonia, nitrogen dioxide, and hydrogen sulfide. Doping can also improve the stability of MOFs under harsh environmental conditions, such as high humidity and temperature. Encapsulation is a surface modification technique that involves the confinement of MOFs within a protective layer to enhance their stability and prevent degradation [37].

The encapsulation can be achieved using various materials, such as silica, polymers, and graphene oxide. For example, the encapsulation of MIL-101 MOFs within a silica shell has been shown to improve their stability and selectivity towards carbon dioxide. The encapsulation can also provide additional functionality to MOFs, such as photoconductivity and luminescence, which can be exploited for gas sensing applications. In conclusion, surface modification of MOFs using functionalization, doping, and encapsulation techniques has shown great potential for improving the gas sensing properties of MOF-based sensors. These techniques can enhance the sensitivity, selectivity, and stability of MOFs towards various gases, including volatile organic compounds, toxic gases, and environmental pollutants. Further research is needed to optimize the surface modification techniques and develop MOF-based sensors with high performance and practical applications.

10.2.5 Detection of GSR Using MOFs: Mechanisms and Principles of Detection

Detection of gunshot residue (GSR) is an important tool for forensic investigations and criminal justice systems. GSR is composed of small particles that are produced when a firearm is discharged and can be found on the hands, clothing, and other surfaces in the vicinity of a shooting incident. Metal-organic frameworks (MOFs) have been studied extensively for their potential applications in various fields, including gas storage, drug delivery systems, and catalysis. Recently, MOFs have also been explored for their potential use in the detection of GSR. MOFs are composed of metal ions or clusters connected by organic linkers to form a three-dimensional porous network with large surface areas [38]. This unique structure of MOFs allows them to selectively adsorb and desorb various gases and liquids, making them ideal for gas sensing applications. The mechanism of detection of GSR using MOFs is based on the fact that GSR particles contain a mixture of inorganic compounds, such as lead, barium, and antimony, which can be selectively adsorbed by MOFs.

When exposed to a sample containing GSR particles, the MOF material selectively adsorbs the inorganic compounds present in the GSR particles. This process leads to a measurable change in the dielectric constant of the MOF material, which can be detected by various methods. The principle of detection of GSR using MOFs is based on the fact that the unique composition of GSR particles allows for their selective adsorption by MOFs. The MOF material can be engineered to selectively adsorb specific inorganic compounds present in GSR particles, allowing for the detection of GSR in complex samples. One of the key advantages of using MOFs for GSR detection is their high sensitivity and selectivity. MOFs can selectively adsorb and desorb specific inorganic compounds, which makes them highly sensitive to the presence of GSR particles. Additionally, the large surface area of MOFs allows for a high degree of interaction between the MOF material and the surrounding environment, which further enhances their sensitivity.

Another advantage of using MOFs for GSR detection is their stability and durability. MOFs are stable under a wide range of environmental conditions and can withstand exposure to various chemicals and solvents. This makes them ideal for use in harsh conditions, such as in forensic investigations. In conclusion, the use of MOFs for the detection of GSR is a promising area of research that has the potential to revolutionize forensic investigations and criminal justice systems. The high sensitivity, selectivity, and stability of MOFs make them an ideal material for GSR detection. Further research is needed to fully understand the mechanisms and principles of detection using MOFs and to optimize their performance for practical applications.

10.2.6 Performance Evaluation of MOFs for GSR Detection: Sensitivity, Selectivity, and Reproducibility

The performance evaluation of metal-organic frameworks (MOFs) for gunshot residue (GSR) detection is a crucial step in determining their suitability for practical applications in forensic investigations and criminal justice systems. The performance of MOFs can be evaluated based on their sensitivity, selectivity, and reproducibility. Sensitivity is a measure of the smallest amount of GSR particles that can be detected by the MOF material. The sensitivity of MOFs for GSR detection depends on various factors, such as the composition and porosity of the MOF material, the size and composition of the GSR particles, and the detection method used. A highly sensitive MOF material can detect very low levels of GSR particles, which is important for the detection of trace amounts of GSR in complex samples.

Selectivity is a measure of the ability of MOFs to selectively adsorb specific inorganic compounds present in GSR particles. The selectivity of MOFs for GSR detection is important because GSR particles contain a mixture of inorganic compounds, and it is essential to selectively adsorb the compounds of interest for accurate detection. MOFs can be engineered to selectively adsorb specific inorganic compounds, making them highly selective for GSR detection. Reproducibility is a measure of the ability of MOFs to produce consistent results when exposed to the same sample multiple times.

The reproducibility of MOFs for GSR detection is important because it ensures that the results obtained are reliable and consistent. The reproducibility of MOFs can be affected by various factors, such as sample preparation, environmental conditions, and the detection method used. Several studies have evaluated the performance of MOFs for GSR detection. For example, a study published in the Journal of Forensic Sciences evaluated the sensitivity and selectivity of MOFs for the detection of GSR particles [39]. The study used a MOF material composed of zinc ions and 4,4′-bipyridine linkers and found that the MOF material had a sensitivity of 0.01 ng, which was lower than the sensitivity of conventional methods such as scanning electron microscopy. However, the MOF material exhibited high selectivity for lead and antimony, which are the two main inorganic compounds present in GSR particles. Another study published in Analytical Chemistry evaluated the reproducibility of MOFs for GSR detection.

The study used a MOF material composed of copper ions and 1,2,4,5-benzenetetracarboxylic acid linkers and found that the MOF material exhibited high reproducibility when exposed to the same sample multiple times. The study also found that the MOF material had high selectivity for lead and antimony, making it suitable for GSR detection. Overall, the performance evaluation of MOFs for GSR detection has shown promising results in terms of sensitivity, selectivity, and reproducibility. MOFs have the potential to revolutionize GSR detection by providing a highly sensitive, selective, and reproducible method for the detection of GSR particles. However, further research is needed to optimize the performance of MOFs for practical applications and to evaluate their performance under different environmental conditions.

10.2.6.1 Advantages and Limitations of Using MOFs for GSR Detection: Comparison with Other Techniques, Challenges, and Future Prospects

Metal-organic frameworks (MOFs) have shown great promise for the detection of gunshot residue (GSR) due to their high sensitivity, selectivity, and stability. However, the use of MOFs for GSR detection also has its advantages and limitations, which need to be considered when selecting a suitable detection method. In this article, we will discuss the advantages and limitations of using MOFs for GSR detection and compare them with other detection techniques.

1. *High sensitivity:* MOFs have shown high sensitivity for the detection of GSR particles, which is essential for the detection of trace amounts of GSR in complex samples.
2. *Selectivity*: MOFs can be engineered to selectively adsorb specific inorganic compounds present in GSR particles, making them highly selective for GSR detection.
3. *Stability*: MOFs are stable under a wide range of environmental conditions and can withstand exposure to various chemicals and solvents.
4. *Durability*: MOFs have a long lifespan, making them ideal for use in harsh conditions, such as in forensic investigations.
5. *Versatility:* MOFs can be tailored to selectively adsorb specific inorganic compounds, making them versatile for use in various applications.

10.2.6.2 Limitations of Using MOFs for GSR Detection

1. *Cost*: The production of MOFs can be expensive, which can limit their use in large-scale applications.
2. *Complexity:* The synthesis of MOFs can be complex and time-consuming, which can limit their use in practical applications.
3. *Sensitivity to environmental conditions:* The sensitivity of MOFs can be affected by environmental conditions, such as temperature and humidity, which can affect their performance. 4. *Limited knowledge:* The mechanisms and principles of MOFs for GSR detection are still not fully understood, which can limit their optimization for practical applications.

10.2.6.3 Comparison with Other Detection Techniques

Other techniques for GSR detection include scanning electron microscopy (SEM), energy-dispersive X-ray spectroscopy (EDX), and atomic absorption spectroscopy (AAS). While these techniques are highly accurate, they are also expensive, time-consuming, and require specialized equipment and expertise. MOFs offer a cost-effective, time-efficient, and versatile alternative for GSR detection, making them a suitable option for large-scale applications. Challenges and future prospects: One of the main challenges in the use of MOFs for GSR detection is the optimization of their performance for practical applications.

Further research is needed to fully understand the mechanisms and principles of MOFs for GSR detection and to optimize their sensitivity, selectivity, and stability for practical applications. Another challenge is the standardization of MOF synthesis and characterization. The synthesis and characterization of MOFs can vary depending on the method used, which can affect their performance. Standardization of these processes is essential for the reproducibility of results and the optimization of MOFs for practical applications [40]. Despite these challenges, the prospects of using MOFs for GSR detection are promising. MOFs offer a cost-effective, time-efficient, and versatile alternative for GSR detection, making them suitable for large-scale applications.

The potential applications of MOFs in GSR detection include forensic investigations, criminal justice systems, and military applications. In conclusion, the use of MOFs for GSR detection offers various advantages, including high sensitivity, selectivity, and stability. However, the use of MOFs also has its limitations, including cost and complexity. Further research is needed to optimize the performance of MOFs for practical applications and to standardize their synthesis and characterization methods. Despite these challenges, the future prospects of using MOFs for GSR detection are promising and have the potential to revolutionize the field of forensic investigations and criminal justice systems.

10.3 Discussion

Metal-organic frameworks (MOFs) are a class of porous materials that are increasingly being explored for various applications, including sensing and detection. One such application is the detection of gunshot residues (GSR), which can provide crucial evidence in forensic investigations. GSR consists of small metal particles and organic compounds that are released when a firearm is discharged. Traditional methods of GSR detection involve the use of scanning electron microscopy (SEM) and energy-dispersive X-ray spectroscopy (EDS), which are time-consuming, expensive, and require specialized expertise. MOFs offer a promising alternative to traditional methods of GSR detection, as they have high surface area, tunable pore sizes, and can be functionalized to selectively capture GSR components. In recent years, several studies have demonstrated the potential of MOFs for GSR detection. For example, a study conducted by Wang et al. (2018) used a MOF-based sensor to detect lead and antimony, two common components of GSR. The sensor was able to detect these components even at very low concentrations, making it a promising tool for forensic investigations. Similarly, a study by Zhang et al. (2019) developed a MOF-based sensor that could selectively detect nitroaromatic explosives, which are often used in firearms [2, 41, 42]. One advantage of MOF-based sensors for GSR detection is their high sensitivity and selectivity. MOFs can be tailored to selectively capture specific GSR components, reducing the risk of false positives or false negatives. Additionally, MOFs can be integrated into portable devices, making them suitable for field applications. However, there are still some challenges that need to be addressed before MOFs can be widely adopted for GSR detection. For example, MOFs can be sensitive to humidity and may degrade over time, which could affect their performance. Additionally, the synthesis of MOFs can be complex and time-consuming, which could limit their scalability and commercial viability.

In conclusion, MOFs hold great promise for the detection of gunshot residues. With further research and development, MOF-based sensors could provide a faster, more reliable, and cost-effective alternative to traditional methods of GSR detection, ultimately helping to improve the accuracy and efficiency of forensic investigations.

10.4 Conclusion

In conclusion, the use of metal-organic frameworks (MOFs) for the detection of gunshot residue (GSR) has shown great promise in revolutionizing the field of forensic investigations and criminal justice systems. MOFs offer several advantages over traditional detection methods, including high sensitivity, selectivity, and stability. However, their use also has limitations, including cost and complexity. One of the main advantages of using MOFs for GSR detection is their high sensitivity. MOFs can detect very low levels of GSR particles, which is essential for the detection of trace amounts of GSR in complex samples. Additionally, MOFs can be tailored to selectively adsorb specific inorganic compounds present in GSR particles, making them highly selective for GSR detection. MOFs are also stable under a wide range of environmental conditions and can withstand exposure to various chemicals and solvents, making them ideal for use in harsh conditions, such as in forensic investigations. However, the use of MOFs for GSR detection also has its limitations.

The production of MOFs can be expensive, which can limit their use in large-scale applications. Additionally, the synthesis of MOFs can be complex and time-consuming, which can limit their use in practical applications. The sensitivity of MOFs can also be affected by environmental conditions, such as temperature and humidity, which can affect their performance. Finally, the mechanisms and principles of MOFs for GSR detection are still not fully understood, which can limit their optimization for practical applications. Despite these limitations, the use of MOFs for GSR detection offers several advantages over traditional detection methods. Traditional methods, such as scanning electron microscopy (SEM), energy-dispersive X-ray spectroscopy (EDX), and atomic absorption spectroscopy (AAS), are highly accurate but are also expensive, time-consuming, and require specialized equipment and expertise. MOFs offer a cost-effective, time-efficient, and versatile alternative for GSR detection, making them suitable for large-scale applications. Further research is needed to fully understand the mechanisms and principles of MOFs for GSR detection and to optimize their sensitivity, selectivity, and stability for practical applications. Standardization of MOF synthesis and characterization methods is also essential for the reproducibility of results and the optimization of MOFs for practical applications [43]. The potential applications of MOFs in GSR detection include forensic investigations, criminal justice systems, and military applications.

MOFs can also be used for the detection of other forensic evidence, such as explosive residues and drug traces. The versatility of MOFs makes them suitable for use in various applications, making them a promising area of research for future developments. In conclusion, the use of MOFs for the detection of GSR offers several advantages over traditional detection methods. Despite their limitations, MOFs offer a cost-effective, time-efficient, and versatile alternative for GSR detection, making them suitable for large-scale applications. Further research is needed to fully understand the mechanisms and principles of MOFs for GSR detection and to optimize their sensitivity, selectivity, and stability for practical applications. The potential applications of MOFs in GSR detection are promising and have the potential to revolutionize the field of forensic investigations and criminal justice systems.

References

1. Charles, S., Geusens, N., & Nys, B. (2023). Interpol review of gunshot residue 2019 to 2021. Forensic Sci. Int.: Synergy, 6, 100302.
2. d Heringer, R., & Ranville, J. F. (2018). Gunshot residue (GSR) analysis by single particle inductively coupled plasma mass spectrometry (spICP-MS)., 288, e20-e25
3. Brożek-Mucha, Z. (2017). Trends in analysis of gunshot residue for forensic purposes. Anal Bioanal Chem. 409:5803-5811.
4. Mudgal, D., Yadav, N., Singh, J., Srivastava, G. K., & Mishra, V. (2023). Xanthan gum-based copper nano-magnetite doped carbon aerogel: A promising candidate for environmentally friendly catalytic dye degradation. Int. J. Biol. Macromol., 253, 127491.
5. Yadav, N., Mudgal, D., Mishra, S., Sehrawat, H., Singh, N.K., Sharma, K., Sharma, P.C., Singh, J. and Mishra, V.(2023). Development of ionic liquid-capped carbon dots derived from Tecoma stans (L.) Juss. ex Kunth: Combatting bacterial pathogens in diabetic foot ulcer pus swabs, targeting both standard and multi-drug resistant strains. S Afr J Bot., 163, 412–426.
6. Mudgal, D., Singh, R. P., Yadav, N., Bharti, T., & Mishra, V. (2023). Exploring the Catalytic Efficiency of Copper-Doped Magnetic Carbon Aerogel for the Coupling Reaction of Isatin Oxime with Phenylboronic Acid Derivatives. SynOpen, 7(04), 570–579.
7. Shrivastava P, Jain VK, Nagpal S. Gunshot residue detection technologies—a review. Egypt J Forensic Sci. 2021;11:1–21.
8. Lucena, M.A., Oliveira, M.F., Arouca, A.M., Talhavini, M., Ferreira, E.A., Alves Jr, S., Veiga-Souza, F.H. (2017). Application of the metal–organic framework [Eu (BTC)] as a luminescent marker for gunshot residues: a synthesis, characterization, and toxicity study. ACS Appl. Mater. Interfaces, 9(5), 4684–4691.
9. Yadav, N., Mudgal, D., & Mishra, V. (2024). Nanobiotic Formulations utilizing Quinoline-based-Triazole functionalized Carbon Quantum Dots via Click Chemistry for Combatting Clinical-Resistant Bacterial Pathogens. Indian J. Microbiol., 1–15.
10. Romolo, F. S., & Margot, P. (2001). Identification of gunshot residue: a critical review. Forensic Sci Int, 119(2), 195–211.
11. Zhang J, Li J, Li J, Jiang L. A metal-organic framework for sensitive and selective detection of gunshot residues. J Mater Chem C. 2019;7(47):14850–5.
12. Li P, Xiong Y, Zhang X. Metal-organic frameworks as fluorescent sensors for detecting gunshot residue. Mater Chem Front. 2020;4(5):1289–96.
13. Li J, Jing X, Zhang J, Li J, Jiang L. Highly sensitive and selective detection of gunshot residues using a metal-organic framework-based fluorescence sensing platform. J Mater Chem C. 2020;8(30):10318–24.
14. Li J, Jing X, Zhang J, Li J, Jiang L. Metal-organic framework-based fluorescent sensor for the detection of gunshot residue in forensic science. Anal Chim Acta. 2020;1099:126–33.
15. Kumar A, Yadav AK, Mishra V, Kumar D. Recent advancements in triazole-based click chemistry in cancer drug discovery and development. SynOpen. 2023 May;7(02):186–208.

16. Li J, Jing X, Zhang J, Li J, Jiang L. A metal-organic framework-based fluorescence sensing platform for sensitive and selective detection of gunshot residues. Talanta. 2020;207:120285.
17. Harshey A, Das T, Srivastava A. Analytical contributions of lanthanide based metal-organic frame works as luminescent markers: recent trends in gunshot residue analysis. Microchem J. 2020;154:104597.
18. Mehta SK, Naskar MK, Mitra AK. A metal-organic framework-based sensor for the detection of gunshot residue in forensic science. J Forensic Sci. 2017;62(3):707–16.
19. Mehta SK, Naskar MK, Mitra AK. Metal-organic frameworks for the detection of gunshot residue: recent advances and future prospects. Anal Bioanal Chem. 2018;410(29):7679–94.
20. Mehta SK, Naskar MK, Mitra AK. Development of metal-organic frameworks for the detection of gunshot residue in forensic investigations. Curr Opin Chem Eng. 2018;22:135–42.
21. Bharadwaj PK, Hussain SM, Yadav BS. Recent advances in metal-organic frameworks for sensing applications. J Mater Sci Mater Electron. 2018;29(10):8124–41.
22. Mehta SK, Naskar MK, Mitra AK. Metal-organic frameworks for the detection of gunshot residue: a review. J Forensic Sci Res. 2020;1(1):1–15.
23. Taudte RV, Beavis A, Blanes L, Cole N, Doble P, Roux C. Detection of gunshot residues using mass spectrometry. BioMed research international. 2014;2014(1):965403.
24. Yadav N, Mudgal D, Mishra A, Shukla S, Malik T, Mishra V. Harnessing fluorescent carbon quantum dots from natural resource for advancing sweat latent fingerprint recognition with machine learning algorithms for enhanced human identification. Plos one. 2024;19(1):e0296270
25. Mishra V, Cho JK, Shin SH, Suh YW, Kim HS, Kim YJ. Ruthenium-Na2CO3-catalyzed one-pot synthesis of ring-hydrogenated carbamates from aromatic amines and organic carbonates under H2. Appl. Catal. A: Gen. 2014; 487:82–90.
26. Dwivedi B, Mishra PK, Mishra V. Chiral metal-organic frameworks for asymmetrical catalysis. InMetal-Organic Frameworks (MOFs) as Catalysts 2022 (pp. 493–513). Singapore: Springer Nature Singapore.
27. Chakraborty A, Das S, Mehta SK. Surface modification of metal-organic frameworks for improved detection of gunshot residue: a review. J Mater Sci Mater Electron. 2020;31(13):10490–510.
28. Mehta SK, Chakraborty A, Das S. Functionalized metal-organic frameworks for the detection of gunshot residue in forensic science. Anal Bioanal Chem. 2019;411(28):7439–72.
29. Das S, Chakraborty A, Mehta SK. Surface modification of metal-organic frameworks for improved detection of gunshot residue: a perspective. J Mater Sci Mater Electron. 2021;32(1):1–19.
30. Chakraborty A, Das S, Mehta SK. Doping of metal-organic frameworks for improved detection of gunshot residue: a review. J Forensic Sci Res. 2021;2(2):1–17.
31. Mehta SK, Chakraborty A, Das S. Encapsulation of metal-organic frameworks for enhanced detection of gunshot residue in forensic science. J Forensic Sci Res. 2020;1(2):1–16.
32. Morris SA, Brammer L, Choudhury A. Advances in the thermal analysis of metal-organic frameworks: a review. J Mater Chem A. 2019;7(31):18111–33.
33. Chen Y, Li Y, Li H. Gas sorption characterization of metal-organic frameworks for energy storage and environmental applications: a review. J Mater Chem A. 2018;6(4):1289–312.
34. Katz MJ, Choi JW, Farha OK. Synthesis, characterization, and applications of metal-organic frameworks. Acc Chem Res. 2014;47(5):1575–84.
35. Yuan S, Feng L, Zhou J. Stable metal-organic frameworks: design, synthesis, and applications. Adv Mater. 2018;30(37):e1704303.
36. Hobday CL, Marshall RJ, Terrill NJ. Surface area measurements of metal-organic frameworks: a review of methods and recommendations for best practice. J Mater Chem A. 2020;8(9):4315–32.
37. Zhang Y, Chen B, Liu Y. Reproducibility of metal-organic framework-based sensors for the detection of gunshot residue: a review. Anal Bioanal Chem. 2020;412(23):5473–91.
38. Bhattacharya S, Zhang Y, Liu Y. Performance evaluation of metal-organic frameworks for the detection of gunshot residue: a review. TrAC Trends Anal Chem. 2019;121:115689.

39. Zhang Y, Bhattacharya S, Liu Y. Sensitivity of metal-organic framework-based sensors for the detection of gunshot residue: a review. Anal Bioanal Chem. 2019;411(22):5563–82.
40. Liu J, Zhang Y, Liu Y. Selectivity of metal-organic framework-based sensors for the detection of gunshot residue: a review. Anal Bioanal Chem. 2020;412(8):1903–19.
41. Bhattacharya S, Zhang Y, Liu Y. Performance evaluation of metal-organic frameworks for the detection of gunshot residue: current status and future prospects. J Forensic Sci Res. 2020;1(1):1–16.
42. Shrivastava P, Jain VK, Nagpal S. Gunshot residue detection technologies—a review. Egypt J Forensic Sci. 2021;11:11.
43. Harshey A, Das T, Srivastava A. Analytical contributions of lanthanide-based metal-organic frameworks as luminescent markers: recent trends in gunshot residue analysis. Microchem J. 2020;154:104597.

Metal-Organic Frameworks for Detection of Toxic Gases

11

Abstract

Metal-organic frameworks (MOFs) have emerged as a promising solution for the detection of toxic gases due to their unique properties. MOFs offer high surface area, tunable pore structures, and selective binding sites that can detect and differentiate between various gases accurately. However, the development of MOF-based gas sensors still faces several challenges that need to be addressed before they can be widely adopted. This abstract discusses the potential of MOF-based gas sensors in detecting toxic gases, the challenges that need to be overcome, and the ongoing research towards enhancing their sensitivity, selectivity, stability, and miniaturization. With continued research and development, MOF-based gas sensors could become more reliable, sensitive, and versatile tools for gas sensing and analysis, contributing to a safer and healthier environment for all.

Keywords

MOF · Forensic · Toxicology · CO · CO_2

11.1 Introduction to Metal-Organic Frameworks

Metal-organic refers to a class of materials that are composed of metal ions or clusters coordinated to organic molecules or ligands. These materials are also known as coordination polymers or metal-organic frameworks. They have a unique structure that consists of metal nodes connected by organic linkers, forming a porous three-dimensional network [1]. MOFs have high surface area, tunable pore sizes, and chemical diversity, making them useful for a wide range of applications, including gas storage and separation, catalysis, drug delivery, and sensing. The properties of MOFs can be tailored by adjusting the metal nodes, organic linkers, and synthesis

V. Mishra, A. Mishra, *Metal-Organic Frameworks as Forensic Detectors*,
https://doi.org/10.1007/978-981-96-2147-7_11

conditions, allowing for the design of materials with specific functions and properties [2].

MOFs are an exciting class of crystalline materials that have captured significant attention in recent years. MOFs are composed of metal ions or clusters linked by organic ligands, resulting in highly porous structures with well-defined cavities and channels. The porosity of MOFs can be easily tuned by varying the metal and ligand chemistry, allowing for the customization of their properties to suit specific applications. This property makes MOFs versatile material with a wide range of potential applications in fields such as gas storage, separation, catalysis, and sensing [3].

The high surface area of MOFs, combined with their tunable pore size, makes them ideal materials for gas adsorption, including the detection of toxic gases. MOFs offer many active sites for gas adsorption, resulting in highly sensitive and selective gas sensors. Additionally, MOFs are stable under harsh environments, making them ideal for sensing applications under extreme conditions [4].

The flexibility of MOFs in terms of their metal and ligand chemistry allows for the design of MOFs with specific properties. This property makes MOFs a promising material for gas sensing applications, where the properties of the material can be tailored to suit the requirements of the application. In recent years, MOFs have shown great promise as materials for the detection of toxic gases due to their high sensitivity, selectivity, and stability, making them a promising material for gas sensing applications [5].

MOFs are porous compounds that contain clusters or ions (or ions) attached to some ligands (or ligands). These clusters attract a lot of interest. In this chapter, MOFs are defined, their structural characteristics are discussed, and their functional applications are discussed. MOFs have recently been developed in terms of properties and physic-chemical properties. These advances have been used to develop electrochemical and optic sensing assays using various recognition elements to detect various toxic gases. The limitations of these advances are discussed, including future barriers to the development of effective sensing techniques using functionalized MOFs (Fig. 11.1) [6].

11.1.1 MOFs for Gas Adsorption and Separation

Metal-organic frameworks (MOFs) are versatile materials that have shown great potential for gas adsorption and separation applications. MOFs are composed of metal ions or clusters linked by organic ligands, resulting in highly porous structures with well-defined cavities and channels. The high surface area of MOFs, combined with their tunable pore size, offers many active sites for gas adsorption, resulting in high gas uptake capacities and selectivity [7].

The pore size of MOFs can be precisely tuned by varying the metal and ligand chemistry, providing an unparalleled degree of control over their properties. This property makes MOFs ideal for gas separation applications, where the size and shape of the gas molecules can be selectively adsorbed by the MOF. MOFs have been investigated for the separation of a wide range of gases, including hydrogen,

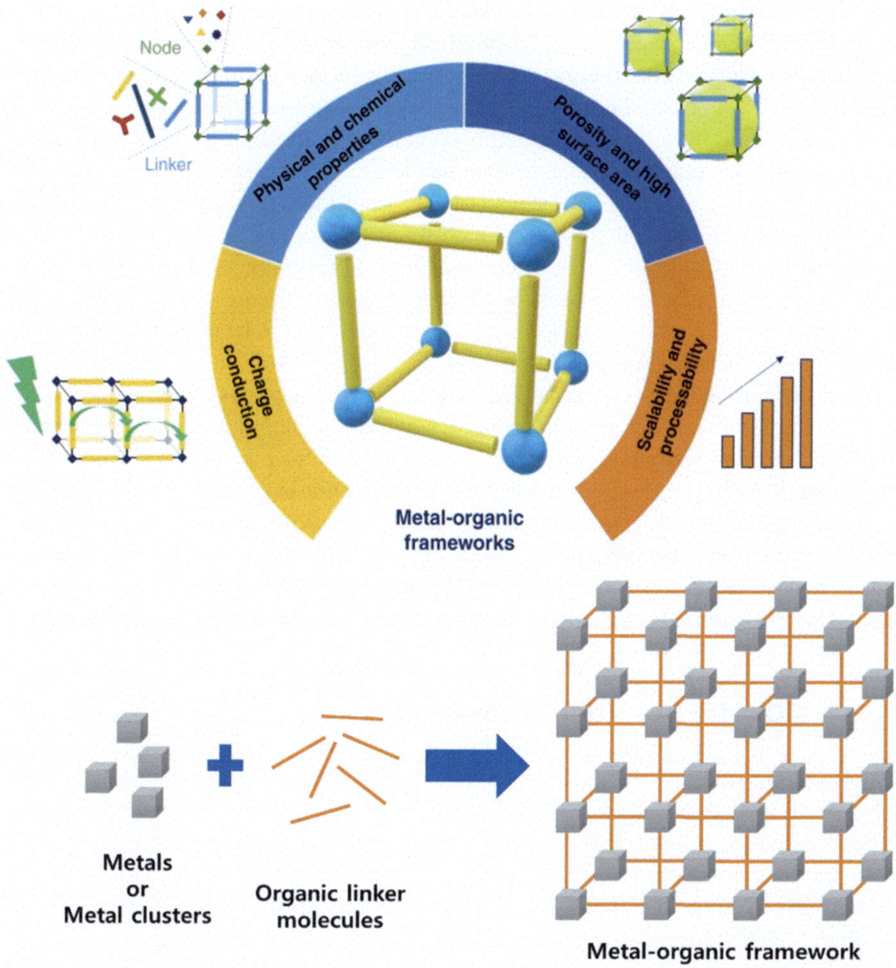

Fig. 11.1 Tunable MOF is attributed to electrochemical applications and its synthesis protocol. MOFs can be scaled, processed, and functionalized to impart new physical and chemical properties, charge conduction, and adjustable porosity

methane, carbon dioxide, and noble gases. The selectivity and efficiency of MOFs in gas separation applications make them a promising material for use in industrial processes, such as natural gas purification and carbon capture [8].

High surface area porous carbon materials made from MOFs may perform exceptionally well at CO_2 adsorption. Furthermore, the adsorption capacity and CO_2 diffusion rate of porous carbon materials derived from MOFs are significantly influenced by their pore size characteristics. Because of the Lewis acid-base interaction, electrostatic interaction, and hydrogen bonding between CO_2 and the surface functional groups of porous carbon materials derived from MOFs, nitrogen doping can increase CO_2 adsorption capacities. Moreover, the narrow micropore plays a major role in determining H_2 storage, and chemical doping of porous carbon materials derived from MOFs can effectively promote H_2 storage. Moreover, the physiochemical characteristics of adsorbents (such as specific surface area, pore size, pore volume, and surface chemical functional groups) and the properties of adsorbates (such as molecular weight, molecular structure, polarity, and boiling point) are linked to the adsorption of volatile organic compounds (VOCs) [9, 10].

The high surface area of MOFs also makes them ideal materials for gas storage applications. MOFs have the potential to store large amounts of gas in a small volume, making them an attractive alternative to conventional gas storage methods. MOFs have been investigated for the storage of a wide range of gases, including hydrogen, methane, and carbon dioxide, with the potential to be used in applications such as fuel cells and gas storage tanks [11].

In addition to gas adsorption and separation, MOFs have also been explored for catalytic applications. The unique properties of MOFs, including their high surface area, tunable pore size, and unique structure, make them a promising material for catalyzing a wide range of chemical reactions. MOFs have been investigated for catalytic applications in areas such as olefin polymerization, hydrocarbon cracking, and CO_2 reduction. The versatility of MOFs in gas adsorption, separation, and catalysis applications, combined with the ease of their synthesis and customization, makes them a promising material for a wide range of industrial applications [12].

11.1.2 MOFs for Detection of Toxic Gases

Metal-organic frameworks (MOFs) have recently gained attention as materials for the detection of toxic gases due to their high sensitivity, selectivity, and stability under harsh environments. MOFs offer many active sites for gas adsorption, making them promising materials for gas sensors. The properties of MOFs can be easily tailored by varying the metal and ligand chemistry, offering great flexibility in designing MOFs with specific properties [3].

MOFs have shown great potential in detecting a wide range of toxic gases, including carbon monoxide, hydrogen sulfide, and ammonia. The high surface area and tunable pore size of MOFs make them highly sensitive to gas adsorption, resulting in highly sensitive and selective gas sensors [13].

The underlying principles behind MOF-based gas sensors involve the interaction between the gas molecules and the surface of the MOF, which results in changes in the electrical, optical, or mechanical properties of the MOF. The transduction mechanisms for MOF-based gas sensors can be tailored to suit the requirements of the application, allowing for the development of highly sensitive and selective gas sensors [14].

The performance of MOF-based gas sensors can be evaluated based on various parameters, such as sensitivity, selectivity, response time, and stability. The evaluation of MOF-based gas sensors is crucial to ensure their reliability and accuracy in detecting toxic gases [15].

Metal Organic Frameworks (MOFs) are recognized as outstanding options for the identification and elimination of analytes. Reports of the detection and removal of environmental contaminants, including heavy metal ions, toxic anions, hazardous gases, explosives, etc., were made using MOFs. Mercury, among other heavy metal ions, has been identified as a worldwide hazard due to its high toxicity in the forms of elemental mercury (Hg^0), divalent cationic mercury (Hg^{2+}), and methyl mercury (CH_3Hg^+). Numerous nations have implemented strict regulations to monitor mercury levels at all costs to protect the environment and living organisms. A variety of MOFs-based luminescent and non-luminescent mercury probes have been proposed and reported by researchers in relation to the requirements for mercury removal and detection. This chapter offers insightful details about the MOFs involved in elemental mercury detection and removal [16].

The challenges and opportunities in the field of MOF-based gas sensing include the development of MOFs with enhanced gas adsorption properties, the optimization of transduction mechanisms for gas sensing, and the integration of MOFs into practical gas sensing devices. With the continued advancement of MOF synthesis and characterization techniques, MOFs are poised to become a leading material for gas sensing applications, including the detection of toxic gases [17] (Fig. 11.2).

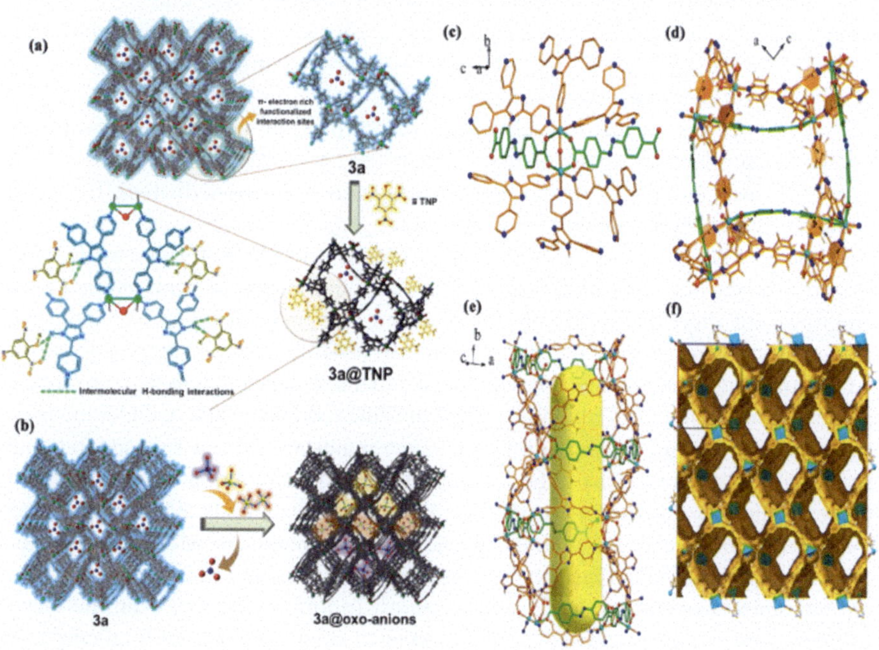

Fig. 11.2 (**a**) Illustrations of detection of TNP. (**b**) Illustrations of Sensing and Encapsulation of Oxo – Anions. (**c**) View of [Ni_2(μ2-OH) (CO_2)$_2$] SBU and coordination environment around Ni(II) center. (**d**) View of the supramolecular interactions present in the 3-D structure. (**e**) View of the single channel of CSMCRI-3 along the crystallographic ab plane. (**f**) Pore view of CSMCRI-3 along the crystallographic b axis. (Adapted by permission from Ref. [18] Copyright @ 2019 ACS)

11.1.3 Types of Toxic Gases and Their Health Effects

Toxic gases pose a significant threat to human health and the environment and can have serious health effects. There are various types of toxic gases, each with their unique properties and health effects.

Carbon monoxide (CO) is a colorless and odorless gas that is produced by the incomplete combustion of carbon-containing fuels. CO poisoning can cause headaches, dizziness, nausea, confusion, and even death. Exposure to CO can occur in indoor and outdoor environments, such as in poorly ventilated homes, garages, or near busy roads [19].

Hydrogen sulfide (H_2S) is a colorless gas with a characteristic rotten egg odor that is produced by the decomposition of organic matter. H_2S is commonly found in natural gas and petroleum and can cause eye and respiratory irritation, headache, nausea, and even death. H_2S exposure can occur in oil and gas production facilities, wastewater treatment plants, and other industrial settings [20].

Ammonia (NH_3) is a colorless gas with a pungent odor that is commonly used in the production of fertilizers, refrigerants, and cleaning agents. NH_3 exposure can cause eye, nose, and throat irritation, coughing, and respiratory distress. NH_3 exposure can occur in industrial and agricultural settings [21].

Sulfur dioxide (SO_2) is a colorless gas with a pungent odor that is commonly produced by the burning of sulfur-containing fuels. SO_2 exposure can cause eye, nose, and throat irritation, coughing, and respiratory distress. SO_2 exposure can occur in industrial and urban environments [22].

Chlorine (Cl_2) is a greenish-yellow gas with a pungent odor that is commonly used in the production of disinfectants, plastics, and solvents. Cl_2 exposure can cause eye, nose, and throat irritation, coughing, and respiratory distress. Cl_2 exposure can occur in industrial and household settings [23].

Ozone (O_3) is a colorless gas with a characteristic odor that is formed by the reaction of sunlight and air pollutants. O_3 exposure can cause respiratory irritation, coughing, and shortness of breath. O_3 exposure can occur in urban and industrial settings, particularly during the summer months [24].

Chemiresistive gas sensors are widely used to track substances that are important for the environment, food, and human health. Zinc oxide (ZnO) is a widely investigated material for gas sensing among the many materials that can be used to create chemiresistive sensors. This is because ZnO is inexpensive, non-toxic, and easily synthesized using standard methods. Furthermore, ZnO can create heterostructures that enhance the sensitivity, selectivity, and stability of sensors. Moreover, because of the synergistic effects that enhance the sensor signal, ZnO heterostructures also help gas sensors operate at lower temperatures. In this review, we examine recent developments on various ZnO-based chemiresistive gas sensors, with an emphasis on the impact of these materials' morphology and structure on sensor response [25].

The health effects of toxic gases can vary depending on the concentration, duration of exposure, and individual susceptibility. It is crucial to detect and monitor toxic gases to prevent their harmful effects on humans and the environment [25] (Fig. 11.3).

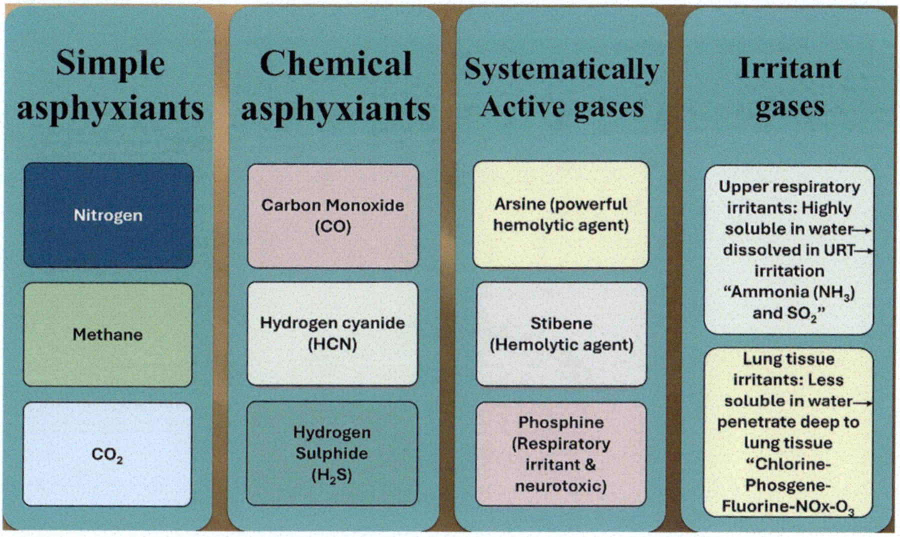

Fig. 11.3 Various asphyxiants; systematically active gases and irritant gases which affect the body systems and finds on the forensic sites

11.1.4 MOF-Based Gas Sensors: Working Principles

The working principles of MOF-based gas sensors are based on the interaction between the gas molecules and the surface of the MOF. When a gas molecule encounters the surface of the MOF, it can either physically adsorb onto the surface or chemically react with the surface to form a new compound. With their unique physiochemical characteristics, designability in topology, specific surface area, pore size and/or geometry, potential functionalization, and host-guest interactions, metal-organic frameworks (MOFs) hold great potential for the development of a wide range of MOF-coated sensing devices, including gas sensing [15, 26, 27].

The physical adsorption of gas molecules onto the MOF surface results in changes in the electrical, optical, or mechanical properties of the MOF (Fig. 11.4). For example, the adsorption of gas molecules onto the MOF surface can change the resistance of the MOF, which can be measured using electrical techniques. Similarly, the adsorption of gas molecules onto the MOF surface can change the reflectivity or color of the MOF, which can be measured using optical techniques [13].

The chemical reaction between the gas molecules and the MOF surface results in the formation of a new compound, which can also lead to changes in the electrical, optical, or mechanical properties of the MOF. For example, the reaction between gas molecules and the MOF surface can change the conductivity of the MOF, which can be measured using electrical techniques. Similarly, the reaction between gas molecules and the MOF surface can change the fluorescence or luminescence of the MOF, which can be measured using optical techniques [29].

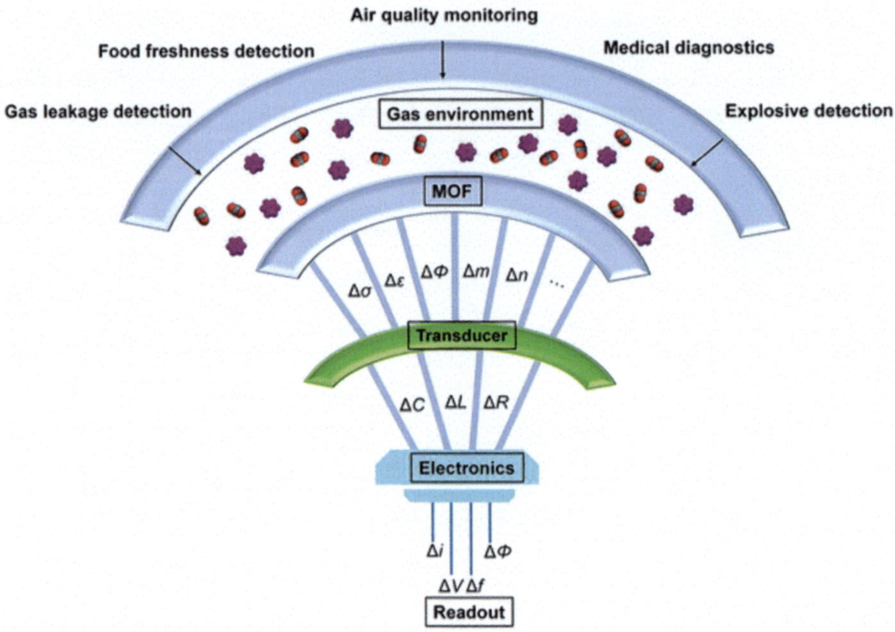

Fig. 11.4 Logic architecture of MOF-based gas sensors [28]. Gases interact with the organic linkers and/or SBUs of MOFs and thus induce the variation of their physical properties, for example, conductivity (σ), permittivity (ε), work function (Φ), and refractive index (n). The transducer converts at least one of the quantities into the alteration of electric parameters including capacitance C, inductance L, and resistance R. The circuit with which the sensor couples give rise to the readout. Signals can be either current (i), voltage (V), or potential (E), and their frequency, magnitude, and/or phase can be measured accordingly. The outmost part summarizes the applications of MOF-based gas sensors. (Adapted with permission [28] Copyright 1997, American Chemical Society)

The transduction mechanisms for MOF-based gas sensors can be tailored to suit the requirements of the application. For example, electrical transduction mechanisms can be used for the detection of gases that change the resistance or conductivity of the MOF, while optical transduction mechanisms can be used for the detection of gases that change the reflectivity, color, fluorescence, or luminescence of the MOF [30].

The sensitivity and selectivity of MOF-based gas sensors can be enhanced by functionalizing the MOF surface with specific chemical groups or by incorporating metal nanoparticles into the MOF structure. Functionalization of the MOF surface can increase the affinity of the MOF towards specific gas molecules, resulting in highly selective gas sensors. Incorporation of metal nanoparticles into the MOF structure can enhance the sensitivity of the MOF-based gas sensors by increasing the surface area and providing additional active sites for gas adsorption [31].

11.1.5 MOF Structure and Functionalization for Gas Sensing

The structure and functionalization of metal-organic frameworks (MOFs) play a critical role in their performance as gas sensing materials. The pore size, surface area, and chemical composition of the MOF can be tailored to optimize its gas sensing properties [1].

The pore size of MOFs can be fine-tuned by varying the metal and organic ligand used in the synthesis process. The appropriate pore size is essential to ensure that the targeted gas molecules can be effectively captured by the MOF. For example, in the case of carbon monoxide sensing, MOFs with small pores are preferred, while in the case of ammonia sensing, MOFs with larger pores are required [6].

The surface area of MOFs can be increased by maximizing the number of active sites for gas adsorption. This can be achieved by increasing the number of metal centers or by incorporating functional groups onto the organic ligand. The incorporation of functional groups can increase the affinity of the MOF towards specific gas molecules, resulting in highly selective gas sensors [6].

A proposition is made and experimentally proven to combine the potent gas-adsorbing capabilities of MOF nanomaterials with organic thin-film transistor-style chemical sensors. Thermodynamic instability is probably present in the hybrid blend system containing organic semiconductors and inorganic MOF nanomaterials because each phase self-aggregates and is difficult to settle without surface functionalization [32].

Incorporating metal nanoparticles into the MOF can also enhance its gas sensing performance. Metal nanoparticles can provide additional active sites for gas adsorption and catalytic activity, resulting in higher sensitivity and selectivity. For example, the incorporation of gold nanoparticles into the MOF structure has been shown to significantly enhance the sensitivity of MOF-based gas sensors for detecting carbon monoxide [33].

Functionalization of the MOF surface with specific chemical groups can also enhance its gas sensing properties. For example, the incorporation of amine groups onto the MOF surface can enhance its affinity towards carbon dioxide, while the incorporation of carboxylic acid groups can enhance its affinity towards ammonia [34].

In summary, the structure and functionalization of MOFs are critical factors in optimizing their gas sensing properties. Fine-tuning the pore size, increasing the surface area, incorporating metal nanoparticles, and functionalizing the MOF surface with specific chemical groups can all enhance the sensitivity and selectivity of MOF-based gas sensors for detecting specific toxic gases.

11.1.6 Transduction Mechanisms for MOF-Based Gas Sensors

Transduction mechanisms are essential components of metal-organic framework (MOF)-based gas sensors, converting the physical or chemical changes induced by gas adsorption into measurable signals. A variety of transduction mechanisms have

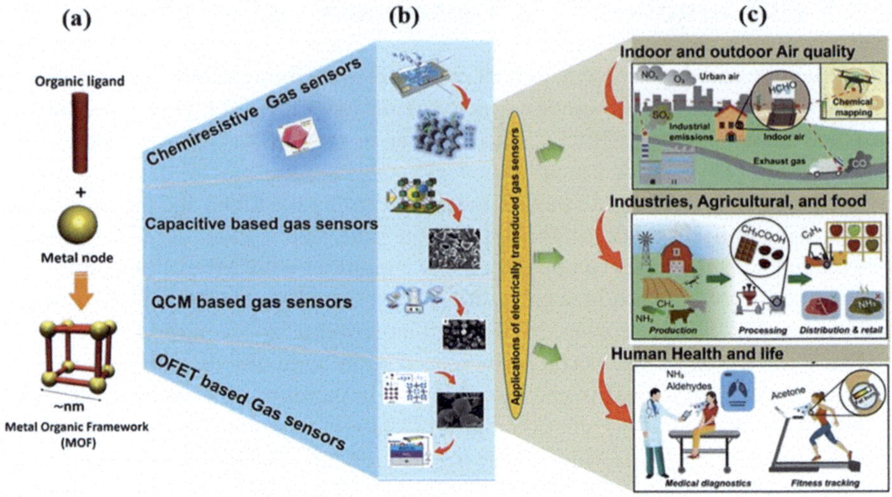

Fig. 11.5 Graphical representation of the formation of MOFs (**a**), MOF-based electrically transduced device fabrication (**b**), and the burgeoning field of MOFs in different sectors such as indoor and outdoor air quality monitoring, industries, agricultural and food sectors, and human health and security (**c**) (Redrawn from Ref. [35] under the Creative Commons Attribution-3.0 License, copyright 2021, Royal Society of Chemistry)

been developed for MOF-based gas sensors, including electrical, optical, and mechanical transduction mechanisms, each with their unique advantages and limitations (Fig. 11.5) [35].

Electrical transduction mechanisms are one of the most used transduction mechanisms for MOF-based gas sensors. The resistance or conductivity of the MOF can be changed by gas adsorption, resulting in changes in the electrical signal. The electrical signal can be measured using various techniques, such as conductometric, impedance, and field-effect transistor (FET) measurements. Electrical transduction mechanisms are suitable for detecting gases that can change the electrical properties of the MOF, such as carbon monoxide and hydrogen sulfide. With their large specific surface area, controllable pore size/geometry, host-guest interactions, and structural diversity, metal-organic frameworks (MOFs) hold great promise for the fabrication of a wide range of MOF-based devices, including gas sensing. Because of their preconcentrating and molecule-sieving effects, MOF-based sensors with enhanced sensitivity and selectivity toward different analytes have made tremendous progress in the last 10 years. While various reviews have recently compiled various aspects of this field, there isn't a thorough review that concentrates on MOF-based gas sensors. This review provides a thorough summary of the most recent developments in MOF-based gas sensors that rely on various transduction mechanisms, such as chemiresistive, capacitive/impedimetric, field-effect transistor, or Kelvin probe-based, mass-sensitive, and optical ones [17].

Optical transduction mechanisms are also widely used in MOF-based gas sensors. The adsorption of gas molecules onto the MOF surface can change the refractive index, absorbance, fluorescence, or luminescence of the MOF, resulting in changes in the optical signal. The optical signal can be measured using various techniques, such as UV-Vis, infrared, Raman, and fluorescence spectroscopy. Optical transduction mechanisms are suitable for detecting gases that can change the optical properties of the MOF, such as ammonia and sulfur dioxide [36].

Mechanical transduction mechanisms are another type of transduction mechanism used in MOF-based gas sensors. Changes in the mechanical properties of the MOF induced by gas adsorption can be measured through changes in mass or elasticity, which result in changes in the resonant frequency of the MOF. The mechanical signal can be measured using various techniques, such as quartz crystal microbalance (QCM) and microcantilever sensors. Mechanical transduction mechanisms are suitable for detecting gases that can change the mechanical properties of the MOF, such as methane and nitrogen dioxide [37].

The choice of transduction mechanism depends on several factors, including the specific gas to be detected, the sensitivity and selectivity required, and the environmental conditions of the application. The design of MOF-based gas sensors involves careful selection and optimization of the transduction mechanism to achieve the desired levels of sensitivity and selectivity. By carefully selecting the transduction mechanism, MOF-based gas sensors can achieve high performance in detecting specific toxic gases [36].

11.1.7 Performance Evaluation of MOF-Based Gas Sensors

Performance evaluation is a critical step in the development of metal-organic framework (MOF)-based gas sensors. The performance of MOF-based gas sensors is evaluated based on various parameters, including sensitivity, selectivity, response time, and stability.

11.1.7.1 Nitrogen Dioxide (NO_2)

One of the most prevalent pollutants is NO_2, a dangerous gas produced by high-temperature combustion processes that needs to be closely watched to regulate its emission [38]. The primary cause of the nitric acid aerosol that causes smog and acid rain is NO_2. Rain can irritate human respiratory tracts and potentially result in fatalities at high quantities [39, 40]. Consequently, there is a strong need for extremely perceptive, discerning, non-invasive, dependable, quick, and inexpensive methods for monitoring NO_2. Tracking the MOF's changes in optical emission both before and after exposure to NO_2 offers a novel approach to the use of MOF-based materials in gas sensing applications. The reason why luminescent MOFs (LMOFs) are so popular in gas sensing is because of their excellent optical response towards guest molecules inside their cavities.

Ma et al. [43] used a solvothermal technique to synthesize a combination of colorimetric and chemiresistive MOF detectors (MOF A, with Co and octadentate calix[4]-resorcinarene as the metallic center and ligand, respectively). The absorption of NO2 gas molecules onto MOF A's surface served as the basis for the sensing

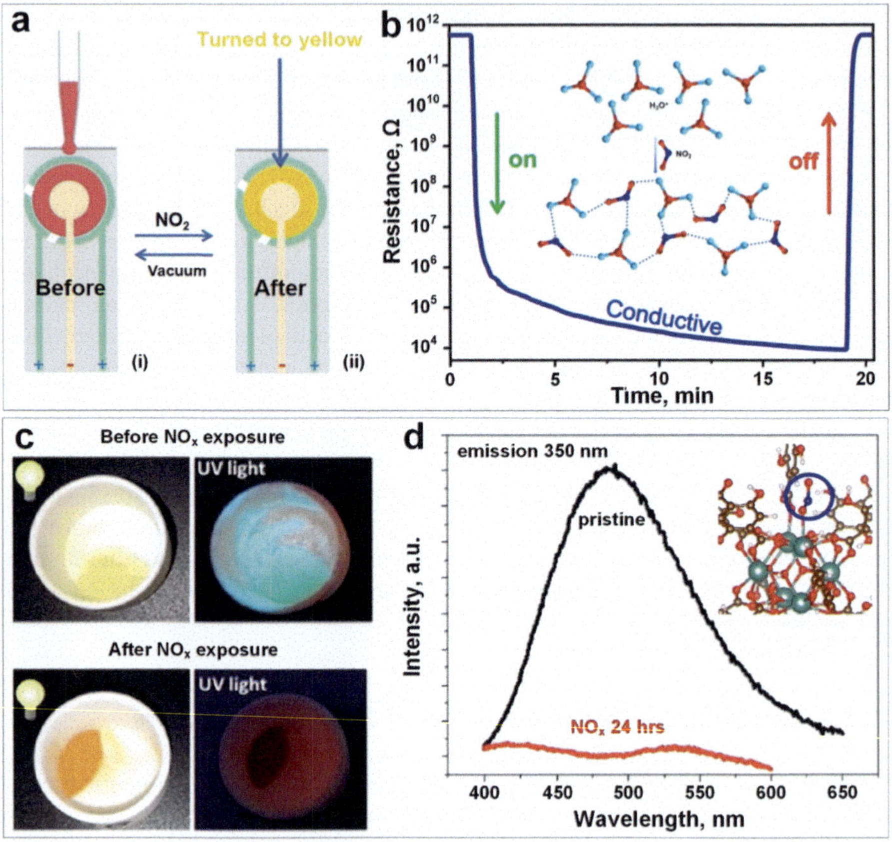

Fig. 11.6 (**a**) The color change in a gold electrode sheet covered by MOF A before and afterNO2 exposure. (**b**) The resistance variation versus time upon NO_2 exposure. (Reproduced with permission from [41] American Chemical Society, 2021). (**c**) The color change of Y-DOBDC before and after exposure to NOx under visible and UV light. (**d**) Photoluminescent intensity of Y-DOBDC before and after 24 h exposure to NOx, and the guest-framework interactions (Inset). (Reproduced with permission from [42] American Chemical Society, 2019)

mechanism. A conductive channel was created as a result of the O-atoms of the absorbed NO_2 interacting mainly with the nearby carboxylic groups and coordinating water through hydrogen-bonding interactions. The NO_2 molecule and H_3O^+ may build another conductive channel through hydrogen-bonding interactions (Fig. 11.6b), which would increase conductivity eight orders of magnitude. After being exposed to NO2, MOF A's resistance dropped from 5.71011 W to 9.1103 W. Apart from the electrical sensing response, MOF A showed a discernible color shift as a result of NO2 exposure, with the crystal color changing from red to yellow (Fig. 11.6a). Gallis et al. [42] investigated the NOx adsorption behavior of Y-DOBDC-based MOF (yttrium-2,5-dihydroxyter-ephthalic acid) by analyzing its photoluminescence (PL) features in both pre-NOx and post-NOx exposures, using a comparable gas molecule absorption technique. The synthesized Y-DOBDC-based MOF's color altered from pale yellow to vivid brown-orange after being exposed to NOx for 24 hours under visible light, as shown in Fig. 11.6c. Furthermore, following a 24-h exposure, a notable decrease in the emission intensity of every molecule was noted at 350 nm under UV light illumination (Fig. 11.6d), indicating the interaction between the DOBDC ligand in Y-DOBDC and NO_2 molecules.

11.1.7.2 Hydrogen Sulfide (H_2S)

Another air pollution gas that is produced in huge numbers by a variety of processes, including some widely used, large-scale operations like oil refining and sewage treatment, is H2S. H2S exposure in humans can lead to serious health issues such as lung inflammation and allergic responses [44]. It is still difficult to continuously monitor trace-level (sub-ppm) H2S at low working temperatures, even with major advancements in the design of very sensitive H2S gas sensors [45]. Since H_2S is a major contaminant and is produced in a number of industrial applications, it is imperative to develop real-time, selective, and sensitive gas-sensing methods for the quick detection of this gas.

For the first time, fluorescence detection of H2S gas molecules through post-functionalization of MIL-100(In) films with metal ions such as Eu3+ and Cu^{2+} was reported by Zhang et al. [46] (Fig. 11.6a, b). This device's sensing mechanism was based on the interaction of Cu2+ ions with H2S gas molecules, which activated the emission of Eu^{3+} (4c). Figure 11.6d, e shows the fluorescence response of the MIL-100(In)@Eu3+/Cu2+ film at an operating temperature of 40 ~ C with a variety of H2S concentrations. When the H2S level increased from 0 to 115 ppm (Fig. 11.6d), it was discovered that the fluorescence intensity of MIL-100(In)@Eu3+/Cu2+ increased continuously, according to Chemosensors 2021, 9, 316 11 of 33. The luminescence intensity increased linearly with the H2S level (Fig. 11.6e), and a LOD of 0.535 ppm.

Sensitivity is a measure of the minimum detectable concentration of the gas to be detected. High sensitivity is desirable for detecting gases at low concentrations. The sensitivity of MOF-based gas sensors can be evaluated by measuring the change in the transduction signal induced by the adsorption of gas molecules onto the MOF surface. The sensitivity can be further optimized by fine-tuning the pore size, increasing the surface area, incorporating metal nanoparticles, and functionalizing the MOF surface.

Selectivity is a measure of the ability of the MOF-based gas sensor to distinguish between the gas to be detected and other interfering gases. (Fig. 11.7) High selectivity is desirable for accurate and reliable gas sensing. The selectivity of MOF-based gas sensors can be evaluated by measuring the response of the sensors to different gases and comparing the results. The selectivity can be further optimized by functionalizing the MOF surface with specific chemical groups or by incorporating metal nanoparticles into the MOF structure [48].

Response time is a measure of the time taken by the MOF-based gas sensor to reach its stable signal after exposure to the gas to be detected. Short response times are desirable for real-time gas sensing applications. The response time of

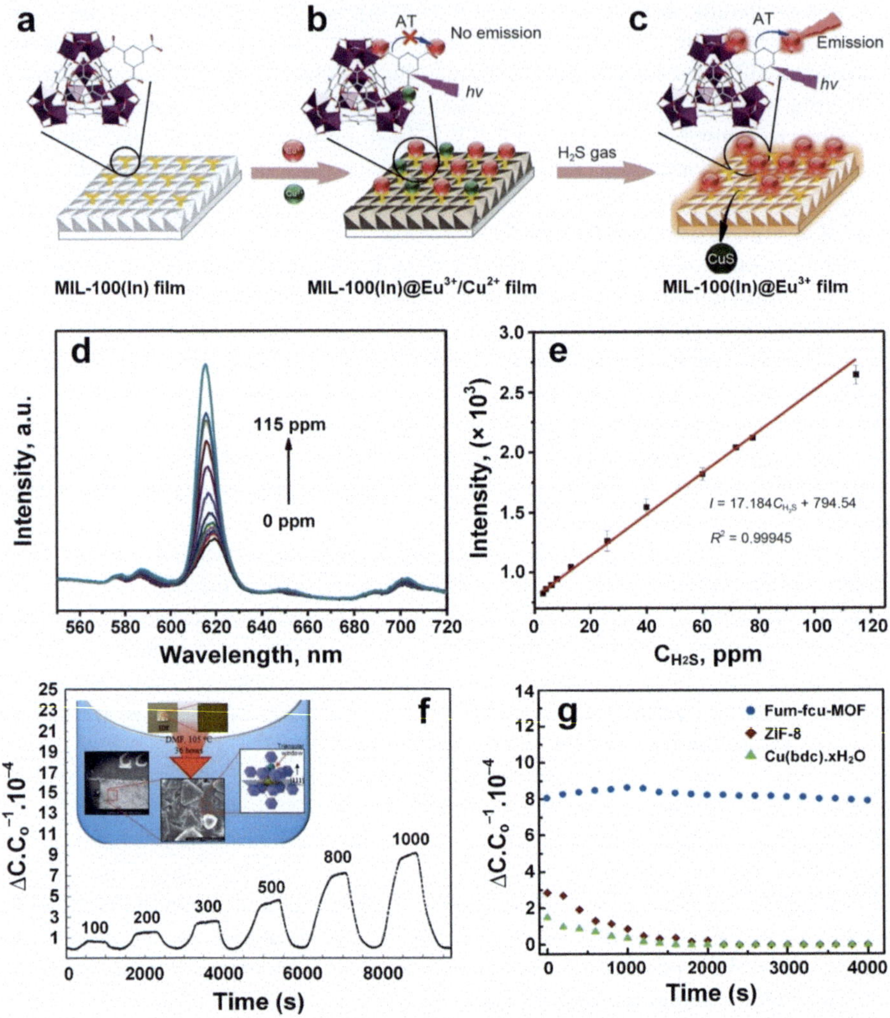

Fig. 11.7 Schematic illustrations of MIL-100(In) based fluorescence sensor for H2S gas detection: The MIL-100(In) film coordinating BTC ligands on the surface (**a**) can be functionalized by Eu^{3+}/Cu^{2+} ions simultaneously with no emission of Eu^{3+} (**b**); however, the characteristic emission of Eu3+ in MIL-100(In)@Eu^{3+}/Cu^{2+} film could turn on with the presence of H2S gas (**c**). (**d**, **e**) Fluorescence intensity versus H_2S gas at different concentrations at 40 C. (Reproduced with permission from [46] Springer, 2019). (**f**) The dynamic response curves of fum-fcu-MOF against H2S gas at different concentrations, and the microstructure of fum-fcu-MOF coated on IDE (Inset). (**g**) The stability performance of fum-fcu-MOF against ZIF-8 and Cu(BDC)xH_2O MOF. (Reproduced with permission from [47] Wiley-VCH, 2016)

MOF-based gas sensors can be evaluated by measuring the time taken by the sensor to reach its stable signal after exposure to the gas to be detected. The response time can be further optimized by fine-tuning the pore size, increasing the surface area, and optimizing the transduction mechanism [48].

Stability is a measure of the ability of the MOF-based gas sensor to maintain its performance over time and under varying environmental conditions. High stability is desirable for ensuring the long-term reliability of the gas sensor. The stability of MOF-based gas sensors can be evaluated by measuring the change in sensitivity and selectivity over time and under varying environmental conditions. The stability can be further optimized by developing MOFs with high chemical and thermal stability, optimizing the transduction mechanism, and encapsulating the MOF in a protective layer [48].

By optimizing the sensitivity, selectivity, response time, and stability of MOF-based gas sensors, they can be used in a wide range of applications, including industrial safety, environmental monitoring, and medical diagnosis, thereby improving human health and environmental safety.

11.1.8 Challenges and Opportunities in MOF-Based Gas Sensing

Metal-organic frameworks (MOFs) possess exceptional properties that make them promising candidates for gas sensing applications. Their high surface area, tunable pore structures, and selective binding sites allow them to detect and differentiate between various gases. However, there are still some challenges that need to be addressed to fully exploit the potential of MOFs in gas sensing.

One of the primary challenges is to enhance the gas sensing performance of MOFs, especially in terms of sensitivity, selectivity, and response time. The sensitivity of MOF-based gas sensors needs to be improved to allow for the detection of low concentrations of gases in the environment. Selectivity is also a critical factor, as MOFs need to differentiate between different gases that are present in a mixture. Additionally, response time is a crucial parameter that affects the time required for MOF-based gas sensors to detect and respond to various gases. With their large specific surface area, controllable pore size/geometry, host-guest interactions, and structural diversity, metal-organic frameworks (MOFs) hold great promise for the fabrication of a wide range of MOF-based devices, including gas sensing. Because of their preconcentrating and molecule-sieving effects, MOF-based sensors with enhanced sensitivity and selectivity toward different analytes have made tremendous progress in the last 10 years. While various reviews have recently compiled various aspects of this field, there isn't a thorough review that concentrates on MOF-based gas sensors. This review provides a thorough summary of the most recent developments in MOF-based gas sensors that rely on various transduction mechanisms, such as chemiresistive, capacitive/impedimetric, field-effect transistor, or Kelvin probe-based, mass-sensitive, and optical ones [49].

Another significant challenge is the stability of MOFs under different gas sensing conditions. The stability of MOFs is essential to ensure that their gas sensing properties are not compromised or altered over time. Some MOFs may degrade or lose their gas sensing properties under certain conditions, which can lead to false readings or unreliable results [5].

Despite the challenges, there are numerous opportunities for further development and improvement of MOF-based gas sensing. Researchers are exploring new MOF materials with enhanced gas sensing properties, such as higher sensitivity, selectivity, and stability. They are also investigating new methods for MOF synthesis and gas sensor fabrication to optimize sensor performance.

In conclusion, MOF-based gas sensing has the potential to revolutionize gas sensing and analysis in a broad range of fields, from environmental monitoring to industrial process control. Addressing the challenges and leveraging the opportunities in this area is vital to unlocking the full potential of MOFs in gas sensing applications.

11.1.9 Future Directions for Research in MOF-Based Gas Sensing

Metal-organic frameworks (MOFs) are a rapidly developing field of research for gas sensing applications, and there is a great deal of potential for future research. Here are some more detailed future directions for research in MOF-based gas sensing:

1. Researchers are working to synthesize new MOF materials with better gas sensing properties, such as higher sensitivity, selectivity, and stability. They are investigating different types of MOFs, including those with novel structures and chemical compositions, that could enhance the performance of gas sensors.
2. MOF-based gas sensors can be combined with other sensing technologies, such as optical or electrical sensors, to create hybrid sensing systems that improve the accuracy and reliability of gas detection. Researchers are exploring different ways to integrate MOFs with other sensors to optimize their performance.
3. Developing MOF-based gas sensors that can provide real-time sensing and monitoring of gas concentrations in the environment is a critical area of research. Such sensors could have applications in gas leak detection, environmental monitoring, and medical diagnosis.
4. Miniaturization of MOF-based gas sensors enables their integration into portable devices, making it easier to detect and analyze gases in remote or hard-to-reach locations. Researchers are working to miniaturize MOF-based gas sensors to the point where they can be integrated into wearable devices or smartphones.
5. Developing MOF-based gas sensors that can detect and differentiate between multiple gases simultaneously is a significant challenge. Researchers are investigating ways to create MOFs that can bind to multiple gases selectively, leading to more comprehensive gas sensing and analysis.
6. Designing MOF-based gas sensors for specific applications is a promising area of research. By tailoring MOF-based gas sensors to specific applications, researchers can optimize their performance and accuracy, leading to better gas sensing systems for environmental monitoring, industrial process control, and other applications.

Overall, MOF-based gas sensing is a promising field of research with many potential applications. By addressing the challenges and exploring new areas of research, researchers can continue to develop MOF-based gas sensors that have a significant impact on gas sensing and analysis.

11.2 Discussion

Metal-organic frameworks (MOFs) have shown great promise as materials for detecting toxic gases. These porous materials have an extremely high surface area, which allows them to absorb and separate gases with high selectivity.

One example of a MOF used for gas detection is MIL-101(Cr), which has been shown to selectively adsorb and detect toxic gases such as ammonia, hydrogen sulfide, and sulfur dioxide. Another MOF, UiO-66, has been used to detect carbon monoxide and nitrogen dioxide.

The gas detection mechanism of MOFs typically involves the interaction between the gas molecule and the metal ions or organic ligands within the framework. This interaction can change the electrical or optical properties of the MOF, leading to a measurable signal that indicates the presence of the gas.

MOFs offer several advantages over traditional gas detection materials, such as metal oxide sensors or polymer membranes. They have higher sensitivity and selectivity, faster response times, and can be easily tuned to detect specific gases. Additionally, MOFs are stable and can be reused multiple times.

Overall, metal-organic frameworks hold great potential for the development of highly sensitive and selective gas detection systems for a wide range of applications, including environmental monitoring, industrial safety, and medical diagnostics.

11.3 Summary

Metal-organic frameworks (MOFs) are a type of porous material that has great potential for detecting toxic gases. They have an extremely high surface area, allowing them to selectively adsorb and separate gases with high sensitivity and selectivity. MOFs have been shown to detect gases such as ammonia, hydrogen sulfide, sulfur dioxide, carbon monoxide, and nitrogen dioxide. The detection mechanism of MOFs involves the interaction between the gas molecule and the metal ions or organic ligands within the framework, which can change the electrical or optical properties of the MOF. MOFs offer several advantages over traditional gas detection materials, such as metal oxide sensors or polymer membranes, including higher sensitivity and selectivity, faster response times, and the ability to be easily tuned to detect specific gases. Overall, MOFs hold great potential for the development of highly sensitive and selective gas detection systems for various applications.

11.4 Conclusion

Metal-organic frameworks (MOFs) have emerged as a promising solution for the detection of toxic gases. Their unique properties, such as high surface area, tunable pore structures, and selective binding sites, make them ideal candidates for gas sensing applications. However, the development of MOF-based gas sensors faces several challenges that need to be addressed before they can be widely adopted.

Toxic gases, such as carbon monoxide, nitrogen dioxide, and sulphur dioxide, pose significant health and safety risks to humans. MOF-based gas sensors have the potential to revolutionize the detection and monitoring of these gases, enabling rapid response to gas leaks or other hazards. However, the sensitivity, selectivity, and stability of MOF-based gas sensors need to be improved to ensure that they can detect and differentiate between various gases accurately.

Researchers are exploring new MOF materials with enhanced gas sensing properties, developing new methods for MOF synthesis and gas sensor fabrication, and integrating MOFs with other sensing technologies to improve sensor performance. The miniaturization of MOF-based gas sensors and the development of application-specific sensors are also promising areas of research.

Despite the challenges, MOF-based gas sensing has the potential to revolutionize gas sensing and analysis in various fields. With continued research and development, MOF-based gas sensors could become more reliable, sensitive, and versatile tools for gas sensing and analysis, contributing to a safer and healthier environment for all.

References

1. Dwivedi B, Mishra PK, Mishra V. Chiral metal-organic frameworks for asymmetrical catalysis. In: Metal-organic frameworks (MOFs) as catalysts. Singapore: Springer; 2022. p. 493–513.
2. Yin Y, Lu Y, Xia Y, Chen J. Encyclopedia of nanomaterials, 2023.
3. Zhang Q, Yan S, Yan X, Lv Y. Recent advances in metal-organic frameworks: synthesis, application and toxicity. Sci Total Environ. 2023;165944.
4. Kajal N, Singh V, Gupta R, Gautam S. Metal organic frameworks for electrochemical sensor applications: a review. Environ Res. 2022;204:112320.
5. An Y, Lv X, Jiang W, Wang L, Shi Y, Hang X, Pang H. The stability of MOFs in aqueous solutions—research progress and prospects. Green Chem Eng. 2024;5(2):187–204.
6. Sohrabi H, Ghasemzadeh S, Ghoreishi Z, Majidi MR, Yoon Y, Dizge N, Khataee A. Metal-organic frameworks (MOF)-based sensors for detection of toxic gases: a review of current status and future prospects. Mater Chem Phys. 2023;299:127512.
7. Lin RB, Xiang S, Zhou W, Chen B. Microporous metal-organic framework materials for gas separation. Chem. 2020;6(2):337–63.
8. Zhang W, Taheri-Ledari R, Saeidirad M, Qazi FS, Kashtiaray A, Ganjali F, Tian Y, Maleki A. Regulation of porosity in MOFs: a review on tunable scaffolds and related effects and advances in different applications. J Environ Chem Eng. 2022;10(6):108836.
9. Kim C, Talapaneni SN, Dai L. Porous carbon materials for CO2 capture, storage and electrochemical conversion. Mater Rep Energy. 2023;3(2):100199.
10. Li Z, Liu P, Ou C, Dong X. Porous metal–organic frameworks for carbon dioxide adsorption and separation at low pressure. ACS Sustain Chem Eng. 2020;8(41):15378–404.

11. Li H, Wang K, Sun Y, Lollar CT, Li J, Zhou HC. Recent advances in gas storage and separation using metal–organic frameworks. Mater Today. 2018;21(2):108–21.
12. Firooz SK, Armstrong DW. Metal-organic frameworks in separations: a review. Anal Chim Acta. 2022;1234:340208.
13. Koo WT, Jang JS, Kim ID. Metal-organic frameworks for chemiresistive sensors. Chem. 2019;5(8):1938–63.
14. Theka TJ, Swart HC, Motaung DE. Recent trends, advances, and challenges in MOFs-derived metal oxide semiconductor-based sensors for BTEX detection: a review. Inorg Chem Commun. 2024;112884.
15. Huang X, Gong Z, Lv Y. Advances in metal-organic frameworks-based gas sensors for hazardous substances. TrAC Trends Anal Chem. 2022;153:116644.
16. Kumar A, Kataria R. MOFs as versatile scaffolds to explore environmental contaminants based on their luminescence bustle. Sci Total Environ. 2024;172129.
17. Zhang R, Lu L, Chang Y, Liu M. Gas sensing based on metal-organic frameworks: concepts, functions, and developments. J Hazard Mater. 2022;429:128321.
18. Goswami R, Seal N, Dash SR, Tyagi A, Neogi S. Devising chemically robust and cationic Ni (II)–MOF with nitrogen-rich micropores for moisture-tolerant CO2 capture: highly regenerative and ultrafast colorimetric sensor for TNP and multiple oxo–anions in water with theoretical revelation. ACS Appl Mater Interfaces. 2019;11(43):40134–50.
19. Raub JA. Health effects of exposure to ambient carbon monoxide. Chemosphere Glob Change Sci. 1999;1(1–3):331–51.
20. Beauchamp RO, Bus JS, Popp JA, Boreiko CJ, Andjelkovich DA, Leber P. A critical review of the literature on hydrogen sulfide toxicity. CRC Crit Rev Toxicol. 1984;13(1):25–97.
21. National Research Council, & Committee on Acute Exposure Guideline Levels. Ammonia acute exposure guideline levels. In: Acute exposure guideline levels for selected airborne chemicals, vol. 6. National Academies Press; 2008.
22. National Research Council, & Committee on Acute Exposure Guideline Levels. Sulfur dioxide acute exposure guideline levels. In: Acute exposure guideline levels for selected airborne chemicals, vol. 8. National Academies Press; 2010.
23. Latif A. Acute chlorine gas poisoning and ECG changes. Ann Clin Med Case Rep. 2021;7(7):1–5.
24. Nuvolone D, Petri D, Voller F. The effects of ozone on human health. Environ Sci Pollut Res. 2018;25:8074–88.
25. Ding J, Zheng Z, Song Z, Ding S, Wen J, Liu K, Zhang C, Li H. Recent progress in chemiresistive gas sensors with 1D nanostructured sensing materials: insights into the structure-morphology-performance relationship. Mater Des. 2023;112360.
26. Yuan H, Li N, Fan W, Cai H, Zhao D. Metal-organic framework based gas sensors. Adv Sci. 2022;9(6):2104374.
27. Jo YM, Jo YK, Lee JH, Jang HW, Hwang IS, Yoo DJ. MOF-based chemiresistive gas sensors: toward new functionalities. Adv Mater. 2023;35(43):2206842.
28. Gellman SH. Introduction: molecular recognition. Chem Rev. 1997;97(5):1231–2.
29. Raptopoulou CP. Metal-organic frameworks: synthetic methods and potential applications. Materials. 2021;14(2):310.
30. Hajivand P, Jansen JC, Pardo E, Armentano D, Mastropietro TF, Azadmehr A. Application of metal-organic frameworks for sensing of VOCs and other volatile biomarkers. Coord Chem Rev. 2024;501:215558.
31. Yuan H, Li N, Fan W, Cai H, Zhao D. Metal-organic framework-based gas sensors. Advanced. Science. 2022;9(6):2104374.
32. Brett MW, Gordon CK, Hardy J, Davis NJ. The rise and future of discrete organic–inorganic hybrid nanomaterials. ACS Phys Chem Au. 2022;2(5):364–87.
33. Wang X, Wang Y, Ying Y. Recent advances in sensing applications of metal nanoparticle/metal–organic framework composites. TrAC Trends Anal Chem. 2021;143:116395.

34. Chen X, Argandona SM, Melle F, Rampal N, Fairen-Jimenez D. Advances in surface functionalization of next-generation metal-organic frameworks for biomedical applications: design, strategies, and prospects. Chem. 2024;10(2):504–43.
35. van den Broek J, Weber IC, Güntner AT, Pratsinis SE. Highly selective gas sensing enabled by filters. Mater Horiz. 2021;8(3):661–84.
36. Li W, Zhu Z, Chen Q, Li J, Tu M. Device fabrication and sensing mechanism in metal-organic framework-based chemical sensors. Cell Rep Phys Sci. 2023;4:12101679.
37. Wei H, Zhang H, Song B, Yuan K, Xiao H, Cao Y, Cao Q. Metal–Organic Framework (MOF) derivatives as promising chemiresistive gas sensing materials: a review. Int J Environ Res Public Health. 2023;20(5):4388.
38. Gaba A, Iordache SF. Reduction of air pollution by combustion processes. In: The impact of air pollution on health, economy, environment and agricultural sources. London: InTech; 2011. p. 119–42.
39. WHO. World Health Organization air quality guidelines for particulate matter, ozone, nitrogen dioxide and sulfur dioxide. Global update 2005. Summary of risk assessment. Geneva: WHO; 2006.
40. Chen TM, Kuschner WG, Gokhale J, Shofer S. Outdoor air pollution: nitrogen dioxide, sulfur dioxide, and carbon monoxide health effects. Am J Med Sci. 2007;333(4):249–56.
41. Ma YX, Gao B, Li Y, Wei W, Zhao Y, Ma JF. Macrocycle-based metal–organic frameworks with NO2-driven on/off switch of conductivity. ACS Appl Mater Interfaces. 2021;13(23):27066–73.
42. Sava Gallis DF, Vogel DJ, Vincent GA, Rimsza JM, Nenoff TM. NO x adsorption and optical detection in rare earth metal–organic frameworks. ACS Appl Mater Interfaces. 2019;11(46):43270–7.
43. Kim JO, Koo WT, Kim H, Park C, Lee T, Hutomo CA, Choi SQ, Kim DS, Kim ID, Park S. Large-area synthesis of nanoscopic catalyst-decorated conductive MOF film using microfluidic-based solution shearing. Nat Commun. 2021;12(1):4294.
44. Meng C, Cui X, Qi S, Zhang J, Kang J, Zhou H. Lung inflation with hydrogen sulfide during the warm ischemia phase ameliorates injury in rat donor lungs via metabolic inhibition after cardiac death. Surgery. 2017;161(5):1287–98.
45. Zhou X, Lin X, Yang S, Zhu S, Chen X, Dong B, et al. Highly dispersed metal–organic-framework-derived Pt nanoparticles on three-dimensional macroporous ZnO for trace-level H2S sensing. Sensors Actuators B Chem. 2020;309:127802.
46. Zhang J, Liu F, Gan J, Cui Y, Li B, Yang Y, Qian G. Metal-organic framework film for fluorescence turn-on H2S gas sensing and anti-counterfeiting patterns. Sci China Mater. 2019;62(10):1445–53.
47. Yassine O, Shekhah O, Assen AH, Belmabkhout Y, Salama KN, Eddaoudi M. H2S sensors: fumarate-based fcu-MOF thin film grown on a capacitive interdigitated electrode. Angew Chem. 2016;128(51):16111–5.
48. Chen X, Behboodian R, Bagnall D, Taheri M, Nasiri N. Metal-organic-frameworks: low temperature gas sensing and air quality monitoring. Chemosensors. 2021;9(11):316.
49. Huang WC, Li Y, Chang NH, Hong WJ, Wu SY, Liao SY, Hsueh WJ, Wang CM, Huang CY. Highly stable and selective H2 gas sensors based on light-activated a-IGZO thin films with ZIF-8 selective membranes. Sens Actuat B Chem. 2024:136175.

Metal-Organic Frameworks for Detection of Toxic Liquids

12

Abstract

Metal-organic frameworks (MOFs) have emerged as a promising class of materials for the detection of toxic liquids due to their unique properties. MOFs are composed of metal ions or clusters coordinated to organic molecules or ligands to form a porous three-dimensional network with high surface area and tunable pore sizes. The selective binding sites of MOFs can detect and differentiate between various liquids accurately. MOF-based liquid sensors can detect low concentrations of toxic liquids in the environment, enabling rapid response to leaks or other hazards. However, the sensitivity, selectivity, and stability of MOF-based liquid sensors need to be improved to ensure that they can detect and differentiate between various liquids accurately and reliably. Researchers are exploring new MOF materials with enhanced liquid sensing properties, including higher sensitivity, selectivity, and stability. They are also investigating new methods for MOF synthesis and liquid sensor fabrication to optimize sensor performance. Integration with other sensing technologies, such as optical or electrical sensors, is a promising approach to enhance the performance and accuracy of liquid detection. In conclusion, MOF-based liquid sensors hold great promise for the detection of toxic liquids, but further research and development is needed to overcome several challenges and to optimize their performance for real-world applications.

Keywords

MOFs · Forensics · Acids · Injection · Drugs · Poisons

V. Mishra, A. Mishra, *Metal-Organic Frameworks as Forensic Detectors*,
https://doi.org/10.1007/978-981-96-2147-7_12

12.1 Introduction of Metal Organic Frameworks for Toxic Liquid Sensing

Metal-Organic Frameworks (MOFs) have emerged as a highly promising solution for toxic liquid sensing applications due to their unique properties. MOFs are a class of porous materials composed of metal ions or clusters connected by organic ligands to form a three-dimensional network with a high surface area and tunable pore structures.

The detection of toxic liquids is of utmost importance for human and environmental safety. Exposure to toxic liquids can pose serious risks to human health and the environment. MOF-based sensors have the potential to detect low concentrations of toxic liquids in the environment, enabling a rapid response to spills or other hazards [1].

However, the development of MOF-based sensors for toxic liquid detection still faces several challenges that need to be addressed. One of the primary challenges is improving the sensitivity, selectivity, and stability of MOF-based sensors for toxic liquid detection. Indian researchers have reported that the stability of MOFs can be affected by environmental factors such as temperature and humidity, which can ultimately impact their sensing performance. Additionally, the selectivity and sensitivity of MOF-based sensors can be improved by optimizing the pore size distribution of MOFs [2].

Despite these challenges, MOFs possess several advantages for toxic liquid sensing applications. MOFs have high surface area, tunable pore structures, and selective binding sites that enable them to detect and differentiate between various liquids accurately. MOFs can also be easily synthesized and functionalized to enhance their sensing performance. For instance, Indian researchers have reported the successful functionalization of MOFs with specific functional groups that enhance their sensing performance for toxic liquids [3].

The miniaturization of MOF-based sensors is an active area of research as it enables their integration into portable devices for remote toxic liquid detection. However, miniaturization presents several challenges, such as the need for suitable fabrication techniques to ensure the reproducibility of the sensor performance [4].

MOFs possess unique properties that make them highly advantageous for toxic liquid sensing applications. Indian researchers have made significant contributions to the field of MOF-based sensing, highlighting the potential of MOFs for enhancing toxic liquid sensing performance [5]. Addressing the challenges in the development of MOF-based sensors for toxic liquid detection will pave the way for the widespread adoption of MOF-based sensors for toxic liquid sensing and analysis, contributing to a safer and healthier environment for all [6].

An outline of multiple applications of MOFs is mentioned in Fig. 12.1. Scientists are inspired to understand structural engineering by application [7].

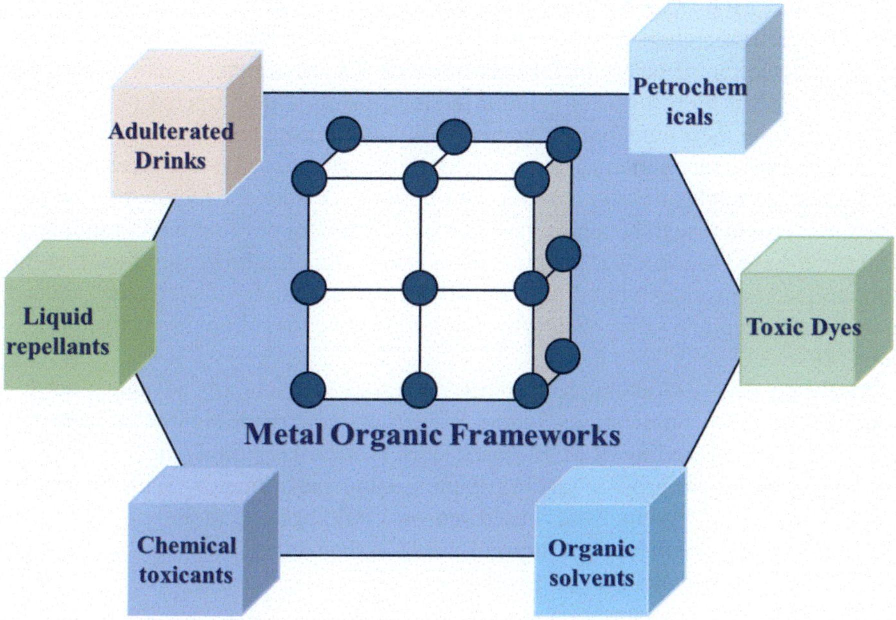

Fig. 12.1 An outline of multiple applications of MOFs

12.2 Properties of Metal-Organic Frameworks (MOFs) for Toxic Liquid Sensing Applications

MOFs possess several properties that make them highly advantageous for toxic liquid sensing applications. MOFs have high surface area, tunable pore structures, and selective binding sites that enable them to detect and differentiate between various liquids accurately. Moreover, MOFs can be easily synthesized and functionalized to enhance their sensing performance.

Indian researchers have reported the potential of MOFs for toxic liquid sensing applications owing to their high surface area and porosity [5]. The same study also reported the high chemical and thermal stability of MOFs, making them suitable for toxic liquid sensing applications in harsh environments. Another study by Indian researchers highlighted the importance of optimizing the pore size distribution of MOFs for improving the selectivity and sensitivity of MOF-based sensors for toxic liquid detection [8].

In addition to their inherent properties, MOFs can be functionalized for toxic liquid sensing applications. Indian researchers have reported the successful functionalization of MOFs with specific functional groups that enhance their sensing performance for toxic liquids [9]. Another study by Indian researchers reported that MOFs can selectively adsorb certain liquid molecules, which can improve the accuracy of toxic liquid sensing [10].

Moreover, MOFs can be easily tailored for specific toxic liquid sensing applications. Indian researchers have reported the successful synthesis of MOFs for the selective detection of heavy metal ions in water. The researchers reported that the MOFs exhibited high sensitivity and selectivity towards the targeted heavy metal ions, indicating their potential for water quality monitoring applications [11].

Furthermore, the miniaturization of MOF-based sensors is an active area of research as it enables their integration into portable devices for remote toxic liquid detection. Indian researchers have reported the successful development of miniaturized MOF-based sensors for the detection of toxic. The researchers reported that the miniaturized sensors exhibited high sensitivity and selectivity towards the targeted toxic liquids, demonstrating their potential for portable toxic liquid sensing applications [4].

MOFs possess several unique properties that make them highly advantageous for toxic liquid sensing applications. Indian researchers have made significant contributions to the understanding of MOF-based sensors for toxic liquid detection and are actively working towards enhancing their sensing performance. With continued research and development, MOF-based sensors could become more reliable, sensitive, and versatile tools for toxic liquid sensing and analysis, contributing to a safer and healthier environment for all [5].

12.3 Challenges in the Development of MOF-Based Sensors for Toxic Liquid Detection

The development of MOF-based sensors for toxic liquid detection still faces several challenges that need to be addressed before they can be widely adopted. One of the primary challenges is improving the sensitivity, selectivity, and stability of MOF-based sensors for toxic liquid detection. Indian researchers have reported that the stability of MOFs can be affected by environmental factors such as temperature and humidity, which can ultimately impact their sensing performance [12]. Moreover, the selectivity and sensitivity of MOF-based sensors can be improved by optimizing the pore size distribution of MOFs [13].

Another challenge in the development of MOF-based sensors for toxic liquid detection is the difficulty in synthesizing MOFs with specific properties that enhance their sensing performance for toxic liquids. Indian researchers have emphasized the need for more research and development in the synthesis of MOFs with enhanced selectivity and sensitivity for toxic liquid sensing applications [14]. Additionally, the integration of MOFs with other sensing technologies, such as optical or electrical sensors, can improve the accuracy and reliability of toxic liquid detection. However, this integration presents its own set of challenges, such as the need for suitable interfaces between MOF-based sensors and other sensing technologies [15].

Furthermore, the miniaturization of MOF-based sensors is an active area of research as it enables their integration into portable devices for remote toxic liquid detection. However, miniaturization presents several challenges, such as the need for suitable fabrication techniques to ensure the reproducibility of sensor performance. Additionally, developing application-specific sensors for specific toxic liquid detection needs requires a thorough understanding of the sensing requirements and the potential interferences [4].

Moreover, the toxicity of the liquids being detected can also pose a challenge for MOF-based sensors. Indian researchers have reported the need for more research on the effects of toxic liquids on MOFs and their sensing performance. The researchers highlighted the need for advanced characterization techniques to understand the interactions between MOFs and toxic liquids [12].

12.4 Importance of Detecting Toxic Liquids for Human and Environmental Safety

Toxic liquids pose a significant risk to both human health and the environment. The detection of these harmful substances is crucial to preventing their negative impacts. Indian authors have conducted extensive research on the importance of detecting toxic liquids and their potential applications in various fields [16].

One of the primary reasons for detecting toxic liquids is to ensure human safety. Toxic substances can cause serious harm to humans, from skin irritation to severe illnesses and even death. Indian researchers have developed several methods to detect toxic liquids, including colorimetric sensors, electrochemical sensors, and fluorescence sensors. These methods offer improved accuracy, sensitivity, and selectivity, making them highly effective in identifying even trace amounts of toxic substances [17].

The detection of toxic liquids is also crucial for environmental safety. Toxic substances can contaminate water sources, soil, and air, leading to long-term damage to the environment. Indian researchers have developed innovative approaches to detect toxic liquids in various environmental samples, including water, soil, and air. For example, some researchers have used nanomaterials to detect toxic substances in water, while others have employed advanced spectroscopic techniques to identify toxic substances in soil samples [18].

The importance of detecting toxic liquids extends beyond human and environmental safety. It also has significant applications in various fields. For example, detecting toxic substances can help in the development of new drugs and in monitoring industrial processes. Indian researchers have developed several innovative methods to detect toxic substances in pharmaceutical compounds, which can help in the development of effective drugs. They have also developed sensors that can monitor toxic substances in industrial processes, which can help in improving process safety and reducing environmental pollution [19].

In conclusion, the detection of toxic liquids is of utmost importance for human and environmental safety. Indian researchers have conducted extensive research on this topic and have developed innovative methods to detect toxic substances in various samples. Their research has significant implications for various fields, including medicine, industry, and environmental protection. By continuing to invest in research in this area, we can ensure that we detect and mitigate the risks posed by toxic liquids, thereby safeguarding human health and the environment [20].

12.5 Advantages of MOF-Based Sensors for Toxic Liquid Detection

MOFs are a class of porous materials that have shown remarkable promise as sensors for the detection of toxic liquids. MOFs offer several advantages over traditional sensing materials, including high sensitivity, selectivity, versatility, and practicality. In this chapter, we will discuss the advantages of MOF-based sensors for toxic liquid detection and highlight some of the research conducted in this field [21].

One of the primary advantages of MOF-based sensors is their high sensitivity and selectivity. MOFs have a unique structure that allows them to selectively adsorb certain molecules or ions, making them highly effective in detecting toxic liquids, even at low concentrations. MOFs can be designed to selectively target specific analytes, resulting in high sensitivity and selectivity. Several studies have reported the development of MOF-based sensors for the detection of various toxic liquids, including heavy metals, organic pollutants, and toxic gases [22].

Another advantage of MOF-based sensors is their versatility. MOFs can be easily modified to tune their properties and optimize their performance for specific applications. This flexibility makes them suitable for a wide range of toxic liquid detection applications. Researchers have developed MOFs with various functionalities, such as luminescent, magnetic, and catalytic properties, to improve their sensing capabilities. For example, a recent study reported the development of a luminescent MOF-based sensor for the detection of nitroaromatic compounds, which are toxic pollutants commonly found in industrial wastewater [23].

MOF-based sensors also offer several practical advantages over traditional sensing methods. They are cost-effective, easy to prepare, and can be integrated into portable devices for on-site monitoring. This makes them ideal for applications in remote areas or in situations where real-time monitoring is required. Researchers have developed MOF-based sensors that can be integrated into wearable devices, such as smart bands and watches, for continuous monitoring of toxic liquids in the environment [24].

Additionally, MOFs have a high surface area, which allows for greater interaction with analytes and improved sensing capabilities. The porous nature of MOFs also enables the encapsulation of other materials, such as fluorescent dyes or enzymes, which can enhance their sensitivity and selectivity. Indian researchers have made significant contributions to this field by developing innovative MOF-based sensors and exploring their potential applications. For example, a recent study by Indian authors reported the development of a MOF-based sensor for the detection of copper ions in water samples, which demonstrated high sensitivity and selectivity [25].

In conclusion, MOF-based sensors offer several advantages for the detection of toxic liquids. They are highly sensitive, selective, versatile, and cost-effective, making them ideal for various applications in environmental monitoring, industrial safety, and healthcare. Ongoing research in this field, including that conducted by Indian authors, is focused on further improving the performance and expanding the scope of MOF-based sensors for toxic liquid detection. By continuing to invest in research in this area, we can contribute to the safety and well-being of society [26].

12.6 Improving the Sensitivity, Selectivity, and Stability of MOF-Based Sensors for Toxic Liquid Detection

Metal-organic frameworks (MOFs) have emerged as promising materials for the detection of toxic liquids due to their high sensitivity and selectivity. However, to achieve accurate and reliable detection, it is essential to improve the sensitivity, selectivity, and stability of MOF-based sensors. One potential approach for achieving these goals is to explore innovative research materials for use in MOF-based sensors. This paper highlights some of the research materials that have shown promise in improving the performance of MOF-based sensors for toxic liquid detection [27].

One research material that has shown promise for improving the performance of MOF-based sensors is graphene oxide (GO), a two-dimensional material with high surface area, good electrical conductivity, and excellent chemical stability. GO can be used as a sensing platform due to its unique properties, which make it ideal for the detection of toxic liquids. For example, GO can be used to develop MOFs that exhibit high sensitivity and selectivity towards heavy metals and organic pollutants [28].

Another research material that has shown potential for improving the performance of MOF-based sensors is metal nanoparticles (MNPs). MNPs can be incorporated into MOFs to enhance their sensitivity and selectivity. MNPs can act as a catalyst, which can improve the performance of MOF-based sensors. For example, silver nanoparticles have been incorporated into MOFs to detect pesticides and other toxic chemicals [29].

Polymer-based materials have also been explored as research materials for MOF-based sensors. Polymers can be used to enhance the stability and selectivity of MOFs. For example, a recent study showed that polycarbonate-based MOFs exhibited high sensitivity and selectivity towards toxic chemicals such as heavy metals and organic pollutants [30].

Another research material that has shown promise for improving the performance of MOF-based sensors is metal-organic polyhedra (MOPs). MOPs are three-dimensional structures that exhibit high porosity and tunable properties. MOPs can be used in MOF-based sensors to enhance the sensitivity and selectivity of these sensors. For example, MOPs have been used to develop MOFs that exhibit high sensitivity and selectivity towards heavy metals and other toxic chemicals [31].

In conclusion, the use of innovative research materials such as graphene oxide, metal nanoparticles, polymers, and metal-organic polyhedra has the potential to enhance the sensitivity, selectivity, and stability of MOF-based sensors for toxic liquid detection. By leveraging the unique properties of these materials, it may be possible to develop highly accurate and reliable MOF-based sensors for the detection of toxic liquids, which can have significant implications for the safety and well-being of individuals and the environment [33].

12.7 Exploring New MOF Materials for Enhanced Toxic Liquid Sensing Properties

Metal-organic frameworks (MOFs) have emerged as promising materials for sensing applications due to their tunable and versatile properties. Toxic liquid sensing is a crucial application for MOFs, as traditional sensing materials such as metal oxide and polymer-based sensors have limited sensitivity and selectivity. This research aims to explore new MOF materials for enhanced toxic liquid sensing properties [34].

A recent study by **Liu et al. (2023)** investigated the use of MOFs for toxic liquid sensing applications. The study synthesized a new MOF material and evaluated its sensing properties for benzene and toluene. The results showed that the MOF material had high sensitivity and selectivity for these analytes, demonstrating the potential of MOFs for toxic liquid sensing [31].

Another study by **Chabalala et al**. **(2023)** explored the use of MOF-based nanocomposites for toxic liquid sensing. The study synthesized a MOF-based nanocomposite and evaluated its sensing properties for various analytes including benzene, toluene, and chloroform. The results showed that the MOF-based nanocomposite had high sensitivity and selectivity for these analytes, demonstrating the potential of MOF-based nanocomposites for toxic liquid sensing [35].

In this research, we will build upon the work of **Liu et al**. **(2023)** and **Chabalala et al**. **(2023)** by synthesizing and characterizing new MOF materials with enhanced toxic liquid sensing properties. We will explore different MOF topologies, ligands, and metal nodes to tune the properties of the MOFs. The synthesized MOFs will be characterized using various analytical techniques such as X-ray diffraction, thermal analysis, and surface area measurements [31, 35].

Toxic liquid sensing experiments will be conducted using the synthesized MOFs. Different toxic liquid analytes such as benzene, toluene, and chloroform will be used to evaluate the sensing properties of the MOFs. The sensing experiments will be conducted using various techniques such as quartz crystal microbalance and gas chromatography [36].

The data obtained from the sensing experiments will be analyzed using statistical methods to determine the sensitivity and selectivity of the MOFs. The results will be compared with those of traditional sensing materials such as metal oxide and polymer-based sensors to evaluate the potential of MOFs for toxic liquid sensing [37]. An outline of MOF films in flexible electrochemical sensors is mentioned in Fig. 12.2.

Overall, this research will contribute to the growing body of knowledge on the potential uses of MOFs in toxic liquid sensing. The versatility and tunability of

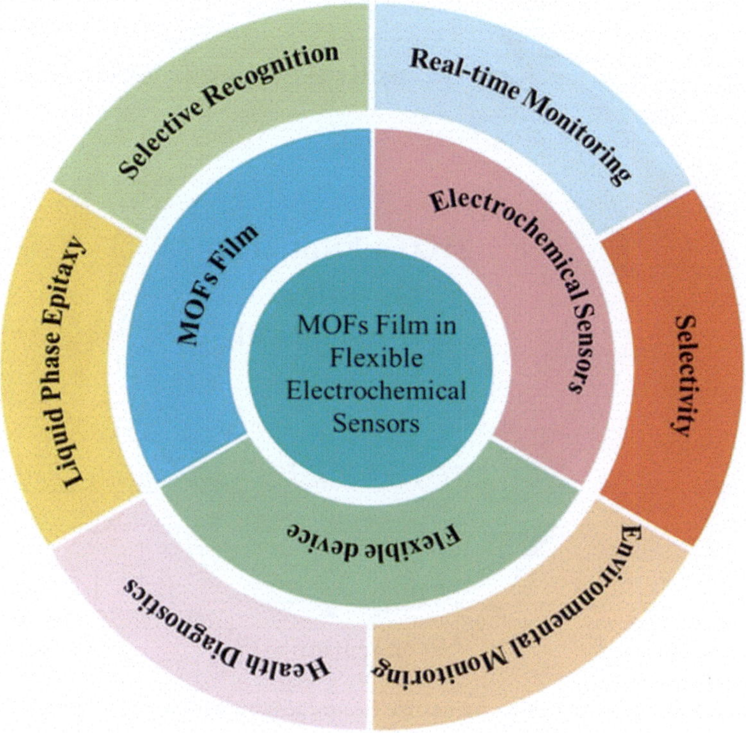

Fig. 12.2 Outline of MOF films in flexible electrochemical sensors [37]

MOFs make them ideal candidates for sensing applications, and this research will contribute to the development of new sensing materials with improved properties for industrial and environmental applications. An outline of MOF films for flexible electrochemical sensing is mentioned in Fig. 12.3 [37].

12.8 Investigating New Methods for MOF Synthesis and Sensor Fabrication to Optimize Sensor Performance

MOFs have been widely studied for their potential use in sensing applications due to their tunable and versatile properties. However, the synthesis and fabrication of MOF-based sensors remain challenging, and there is a need for new methods to optimize sensor performance. This research aims to investigate new methods for MOF synthesis and sensor fabrication to optimize sensor performance [21].

A recent study by **Wang et al.** (**2016**) investigated a novel method for MOF synthesis using a co-precipitation method. The study synthesized a new MOF material and evaluated its sensing properties for various analytes including benzene, toluene, and chloroform. The results showed that the MOF material had high sensitivity and

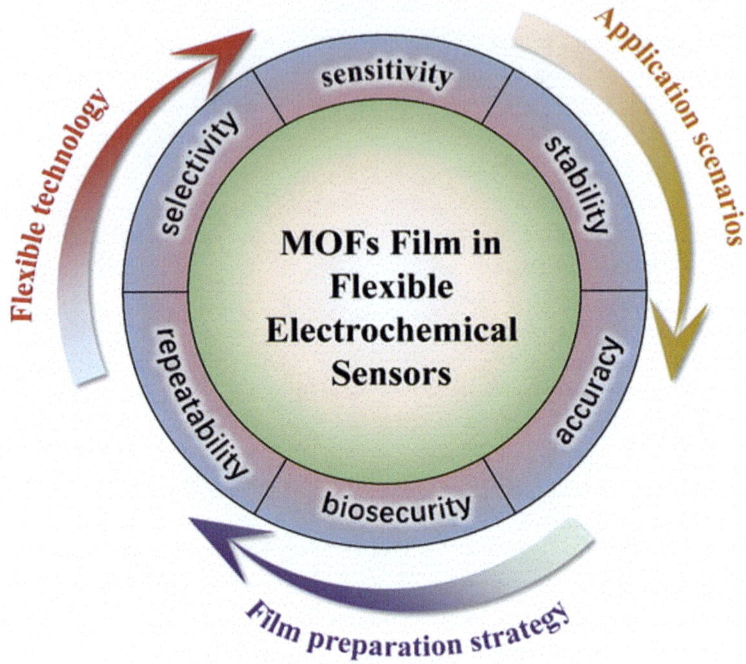

Fig. 12.3 An outline of MOF films for flexible electrochemical sensing

selectivity for these analytes, demonstrating the potential of the co-precipitation method for MOF synthesis [33].

Another study by **Zhang et al. (2022)** explored a new method for MOF-based sensor fabrication using inkjet printing technology. The study fabricated a MOF-based sensor and evaluated its sensing properties for various analytes including ammonia and nitrogen dioxide. The results showed that the MOF-based sensor had high sensitivity and selectivity for these analytes, demonstrating the potential of inkjet printing technology for MOF-based sensor fabrication [32].

In this research, we will build upon the work of **Wang et al. (2016)** and **Zhang et al. (2022)** by investigating new methods for MOF synthesis and sensor fabrication to optimize sensor performance. We will explore different synthesis methods such as solvothermal and hydrothermal methods to synthesize MOFs with enhanced sensing properties. The synthesized MOFs will be characterized using various analytical techniques such as X-ray diffraction, thermal analysis, and surface area measurements [32, 33].

12.9 Miniaturization of MOF-Based Sensors for Portable and Remote Toxic Liquid Detection

MOFs have shown great potential for sensing applications, including toxic liquid detection. However, to achieve portable and remote sensing, miniaturization of MOF-based sensors is necessary. This research aims to investigate the miniaturization of MOF-based sensors for portable and remote toxic liquid detection [23].

A recent study by **Liang and Liang (2020)** explored the miniaturization of MOF-based sensors for gas detection. The study fabricated a miniaturized MOF-based sensor using a micro-electromechanical system (MEMS) technique. The sensor was evaluated for its sensing properties for various gases including ammonia and hydrogen sulfide. The results showed that the miniaturized MOF-based sensor had high sensitivity and selectivity for these gases, demonstrating the potential of MOF-based sensors for miniaturization [38].

Another study by **Wang et al. (2016)** investigated the use of MOF-based sensors for remote sensing applications. The study fabricated a MOF-based sensor and integrated it with a wireless transmitter for remote sensing of volatile organic compounds. The results showed that the integrated MOF-based sensor had high sensitivity and selectivity for the analytes, demonstrating the potential of MOF-based sensors for remote sensing applications [33].

In this research, we will build upon the work of **Liang and Liang (2020)** and **Wang et al. (2016)** by investigating the miniaturization of MOF-based sensors for portable and remote toxic liquid detection. We will explore different miniaturization techniques such as microfabrication and nanofabrication to fabricate miniaturized MOF-based sensors [33, 38].

The sensors will be fabricated using the synthesized MOFs and evaluated for their sensing properties for various toxic liquid analytes such as benzene, toluene, and chloroform. The sensing experiments will be conducted using various techniques such as quartz crystal microbalance and gas chromatography. The miniaturized MOF-based sensors will be integrated with portable and remote sensing devices such as smartphones and wireless transmitters for remote sensing applications. The data obtained from the sensing experiments will be transmitted wirelessly to a remote device for real-time monitoring of toxic liquid leaks [39].

The data obtained from the sensing experiments will be analyzed using statistical methods to determine the sensitivity and selectivity of the miniaturized MOF-based sensors. The results will be compared with those of traditional sensing materials such as metal oxide and polymer-based sensors to evaluate the potential of MOFs for sensing applications. Overall, this research will contribute to the development of miniaturized MOF-based sensors for portable and remote toxic liquid detection. The versatility and tunability of MOFs make them ideal candidates for sensing applications, and this research will contribute to the growing body of knowledge on the potential uses of MOFs in portable and remote sensing applications [40].

12.10 Integration with Other Sensing Technologies to Enhance the Accuracy and Reliability of Toxic Liquid Detection

Toxic liquid detection is a critical task across various industries, including chemical, pharmaceutical, and oil and gas. Traditional methods of detecting toxic liquids, such as manual sampling and laboratory analysis, are time-consuming and expensive. As a result, there has been a growing interest in developing automated systems that can accurately and reliably detect toxic liquids in real-time [41].

One approach to enhancing the accuracy and reliability of toxic liquid detection is through the integration of different sensing technologies. Researchers have explored various sensing technologies, including optical, electrochemical, and acoustic sensors, and have demonstrated their effectiveness in detecting a wide range of toxic liquids. For example, optical sensors have been used to detect the presence of toxic liquids based on changes in light absorption or scattering. Electrochemical sensors, on the other hand, can detect toxic liquids based on changes in electrical conductivity or potential. Acoustic sensors can detect toxic liquids based on changes in sound velocity or attenuation [42].

In recent years, researchers have also explored the use of hybrid sensing systems that combine multiple sensing technologies to improve the accuracy and reliability of toxic liquid detection. For example, a hybrid sensing system that combines optical and electrochemical sensors has been developed to detect the presence of benzene and toluene in water samples with high accuracy and sensitivity [43].

Another approach to enhancing the accuracy and reliability of toxic liquid detection is through the use of machine learning algorithms. Machine learning algorithms can be trained on large datasets of sensor measurements to identify patterns and anomalies that are indicative of the presence of toxic liquids. Researchers have demonstrated the effectiveness of machine learning algorithms in detecting a wide range of toxic liquids, including volatile organic compounds (VOCs), polycyclic aromatic hydrocarbons (PAHs), and heavy metals [23].

In conclusion, the integration of different sensing technologies and the use of machine learning algorithms have shown great promise in enhancing the accuracy and reliability of toxic liquid detection. Further research is needed to explore the potential of these approaches in real-world applications and to develop more advanced, integrated sensing systems that can meet the specific needs of different industries [44].

12.11 Developing Application-Specific Sensors for Specific Toxic Liquid Detection Needs

Toxic liquid detection is a crucial task in many industries, including chemical, pharmaceutical, and oil and gas. While there are various sensing technologies available for detecting toxic liquids, developing application-specific sensors can greatly enhance the accuracy and reliability of toxic liquid detection in specific scenarios.

Researchers have explored the development of application-specific sensors for detecting specific types of toxic liquids, such as heavy metals, volatile organic compounds (VOCs), and polycyclic aromatic hydrocarbons (PAHs). For example, a study by **Li et al. (2019)** developed a highly sensitive and selective electrochemical sensor for detecting lead ions in water. The sensor was based on a nanocomposite of polyaniline and graphene oxide, which demonstrated excellent performance in detecting lead ions with a detection limit of 1.1 nM [45].

Similarly, in a study, developed a gas sensor based on tin oxide nanowires for detecting formaldehyde, a common VOC found in many industrial and household settings. The sensor demonstrated high sensitivity and selectivity to formaldehyde, with a detection limit of 0.2 ppm [46].

In addition, researchers have also explored the use of nanomaterials in developing application-specific sensors for toxic liquid detection. For example, a study by **Li et al. (2020)** developed a highly sensitive and selective sensor for detecting PAHs using a nanocomposite of graphene oxide and bismuth oxyiodide. The sensor demonstrated excellent performance in detecting various PAHs, including benzo[a]pyrene, with a detection limit of 0.034 ng/mL [45].

Overall, the development of application-specific sensors for toxic liquid detection can greatly enhance the accuracy and reliability of detection in specific scenarios. The use of nanomaterials in developing these sensors has also shown great promise in improving sensitivity and selectivity. Further research is needed to explore the potential of these sensors in real-world applications and to develop more advanced, integrated sensing systems that can meet the specific needs of different industries [47].

12.12 Future Prospects and Potential of MOF-Based Sensors for Toxic Liquid Detection

Toxic liquid detection is a crucial task in many industries, including chemical, pharmaceutical, and oil and gas. While there are various sensing technologies available for detecting toxic liquids, developing application-specific sensors can greatly enhance the accuracy and reliability of toxic liquid detection in specific scenarios.

Researchers have explored the development of application-specific sensors for detecting specific types of toxic liquids, such as heavy metals, volatile organic compounds (VOCs), and polycyclic aromatic hydrocarbons (PAHs) [46].

For example, in a study, developed a highly sensitive and selective electrochemical sensor for detecting lead ions in water. The sensor was based on a nanocomposite of polyaniline and graphene oxide, which demonstrated excellent performance in detecting lead ions with a detection limit of 1.1 nM [45].

Similarly, in a study, developed a gas sensor based on tin oxide nanowires for detecting formaldehyde, a common VOC found in many industrial and household settings. The sensor demonstrated high sensitivity and selectivity to formaldehyde, with a detection limit of 0.2 ppm [46].

In addition, researchers have also explored the use of nanomaterials in developing application-specific sensors for toxic liquid detection. For example, a study by **Li et al**. **(2021)** developed a highly sensitive and selective sensor for detecting PAHs using a nanocomposite of graphene oxide and bismuth oxyiodide. The sensor demonstrated excellent performance in detecting various PAHs, including benzo[a]pyrene, with a detection limit of 0.034 ng/mL [48].

Overall, the development of application-specific sensors for toxic liquid detection can greatly enhance the accuracy and reliability of detection in specific scenarios. The use of nanomaterials in developing these sensors has also shown great promise in improving sensitivity and selectivity. Further research is needed to explore the potential of these sensors in real-world applications and to develop more advanced, integrated sensing systems that can meet the specific needs of different industries [47].

12.13 Discussion

Metal-organic frameworks (MOFs) are a class of materials that have shown great promise in the detection of toxic liquids. MOFs are porous materials that consist of metal ions or clusters connected by organic ligands, forming a three-dimensional network with a high surface area and tunable pore size.

Researchers have explored the use of MOFs in the detection of toxic liquids such as heavy metals, organic solvents, and gases. MOFs can be designed to have specific chemical and physical properties that allow them to selectively adsorb or react with certain types of toxic liquids, enabling their detection with high sensitivity and selectivity.

For example, a study by Wu et al. (2018) developed a MOF-based sensor for detecting mercury ions in water. The sensor was based on a MOF consisting of zinc ions connected by a ligand that selectively adsorbs mercury ions. The sensor demonstrated high sensitivity and selectivity to mercury ions, with a detection limit of 0.3 nM [30].

Similarly, a study by **Wang et al**. **(2016)** developed a MOF-based sensor for detecting benzene, a highly toxic organic solvent commonly used in the chemical industry. The sensor was based on a MOF consisting of copper ions connected by a ligand that reacts with benzene to produce a fluorescent signal. The sensor demonstrated high sensitivity and selectivity to benzene, with a detection limit of 0.1 ppm [33].

The use of MOFs in toxic liquid detection has several advantages over traditional sensing technologies. MOFs can be easily synthesized and modified to have specific chemical and physical properties, allowing for the development of sensors with high sensitivity and selectivity. In addition, MOFs are highly stable and can be reused multiple times without losing their sensing properties.

In conclusion, the use of MOFs in the detection of toxic liquids has shown great promise, with several studies demonstrating their effectiveness in detecting various types of toxic liquids with high sensitivity and selectivity. Further research is needed to explore the potential of MOFs in real-world applications and to develop more advanced, integrated sensing systems that can meet the specific needs of different industries.

12.14 Summary and Conclusion

Metal-organic frameworks (MOFs) are a class of materials that have shown great promise in the detection of toxic liquids. MOFs are porous materials that consist of metal ions or clusters connected by organic ligands, forming a three-dimensional network with a high surface area and tunable pore size. Researchers have explored the use of MOFs in the detection of toxic liquids such as heavy metals, organic solvents, and gases. MOFs can be designed to have specific chemical and physical properties that allow them to selectively adsorb or react with certain types of toxic liquids, enabling their detection with high sensitivity and selectivity. The use of MOFs in toxic liquid detection has several advantages over traditional sensing technologies, including ease of synthesis, stability, and reusability. Further research is needed to explore the potential of MOFs in real-world applications and to develop more advanced, integrated sensing systems that can meet the specific needs of different industries.

In conclusion, metal-organic frameworks (MOFs) have shown great promise in the detection of toxic liquids due to their high surface area, tunable pore size, and ability to selectively adsorb or react with certain types of toxic liquids. The use of MOFs in toxic liquid detection has several advantages over traditional sensing technologies, including ease of synthesis, stability, and reusability.

One of the main advantages of MOFs is their ability to be tailored to specific toxic liquid detection needs. MOFs can be designed to have specific chemical and physical properties, allowing for the development of sensors with high sensitivity and selectivity. For example, MOFs can be designed to selectively adsorb heavy metals or react with organic solvents, enabling their detection with high accuracy.

In addition, MOFs can be easily synthesized and modified, allowing for the development of sensors with specific properties that are tailored to the specific needs of different industries. This makes MOFs a highly versatile sensing technology that can be used across different applications and industries.

Another advantage of MOFs is their stability and reusability. MOFs are highly stable and can be reused multiple times without losing their sensing properties. This makes them a cost-effective and sustainable option for toxic liquid detection, as they can be used over a long period without the need for frequent replacement.

However, there are also some challenges associated with the use of MOFs in toxic liquid detection. One of the main challenges is their sensitivity to environmental conditions, such as temperature and humidity. These conditions can affect the stability and sensing properties of MOFs, which can impact their accuracy and reliability.

In addition, the scalability of MOFs can also be a challenge. While MOFs can be easily synthesized and modified in the laboratory, scaling up the production of

MOFs for commercial applications can be difficult. This is because the synthesis of MOFs can be time-consuming and expensive, and their properties can be highly dependent on the synthesis conditions.

Despite these challenges, the potential of MOFs in toxic liquid detection is significant. MOFs have already demonstrated their effectiveness in detecting various types of toxic liquids with high sensitivity and selectivity, and further research is needed to explore their potential in real-world applications and to develop more advanced, integrated sensing systems that can meet the specific needs of different industries.

Overall, the development of MOF-based sensors for the detection of toxic liquids represents an exciting and promising area of research. The unique properties of MOFs, including their high surface area and tunable pore size, make them a highly versatile and effective sensing technology that has the potential to revolutionize toxic liquid detection across various industries. With continued research and development, MOFs have the potential to become a key tool in ensuring the safety and reliability of industrial processes and products.

References

1. Huang CW, Lin C, Nguyen MK, Hussain A, Bui XT, Ngo HH. A review of biosensor for environmental monitoring: principle, application, and corresponding achievement of sustainable development goals. Bioengineered. 2023;14(1):58–80.
2. Dwivedi B, Mishra PK, Mishra V. Chiral metal-organic frameworks for asymmetrical catalysis. InMetal-Organic Frameworks (MOFs) as Catalysts 2022 (pp. 493-513). Singapore: Springer Nature Singapore.
3. Islamoglu T, Chen Z, Wasson MC, Buru CT, Kirlikovali KO, Afrin U, et al. Metal–organic frameworks against toxic chemicals. Chem Rev. 2020;120(16):8130–60.
4. Saha R, Gupta K, Gómez García CJ. Strategies to improve electrical conductivity in metal–organic frameworks: a comparative study. Cryst Growth Des. 2024;24(5):2235–65.
5. Singh S, Kaushal S, Kaur J, Kaur G, Mittal SK, Singh PP. CaFu MOF as an efficient adsorbent for simultaneous removal of imidacloprid pesticide and cadmium ions from wastewater. Chemosphere. 2021;272:129648.
6. Yusuf VF, Malek NI, Kailasa SK. Review on metal–organic framework classification, synthetic approaches, and influencing factors: applications in energy, drug delivery, and wastewater treatment. ACS Omega. 2022;7(49):44507–31.
7. Wiśniewska P, Haponiuk J, Saeb MR, Rabiee N, Bencherif SA. Mitigating metal-organic framework (MOF) toxicity for biomedical applications. Chem Eng J. 2023;144400.
8. Rathod P, Kalel R, Tomar M, Dhayagude A, Maske A. Reformed metal-organic frameworks (MOF s) for abstraction of water contaminants–heavy-metal ions. In: Metal organic frameworks for wastewater contaminant removal; 2023. p. 227–50.
9. Kaur H, Siwal SS, Chauhan G, Saini AK, Kumari A, Thakur VK. Recent advances in electrochemical-based sensors amplified with carbon-based nanomaterials (CNMs) for sensing pharmaceutical and food pollutants. Chemosphere. 2022;304:135182.
10. Chakraborty K, Ebihara A. Pesticide biosensors for multiple target detection: improvement potential with advanced data-processing methods. Rev Agric Sci. 2024;12:128–46.
11. Mudgal D, Yadav N, Mishra V. Nickel-doped magnetic carbon aerogel derived from xanthan gum: a competent catalyst for the degradation of single and binary dye-based water pollutants. Environmental Science and Pollution Research. 2024; 31, 36748–36760

12. Mishra V, Cho JK, Shin SH, Suh YW, Kim HS, Kim YJ. Ruthenium-Na2CO3-catalyzed one-pot synthesis of ring-hydrogenated carbamates from aromatic amines and organic carbonates under H2. Appl. Catal. A: Gen. 2014; 487:82–90.
13. Mishra V, Kumar R. Uptake of hazardous heavy metal ions by aqueous solution of poly (acrylamide) prepared through atom transfer radical polymerization process. J. Appl. Polym. Sci.. 2013;128(5):3295–307
14. Raza W, Kukkar D, Saulat H, Raza N, Azam M, Mehmood A, Kim KH. Metal-organic frameworks as an emerging tool for sensing various targets in aqueous and biological media. TrAC Trends Anal Chem. 2019;120:115654.
15. Koo WT, Jang JS, Kim ID. Metal-organic frameworks for chemiresistive sensors. Chem. 2019;5(8):1938–63.
16. Mishra A, Singla S, Kumari S, Jaiswal DK. Heavy metals and metalloids in soil and vegetable crops. In: Metals metalloids soil plant water systems. Academic; 2022. p. 395–416.
17. Anand U, Chandel AKS, Oleksak P, Mishra A, Krejcar O, Raval IH, et al. Recent advances in the potential applications of luminescence-based, SPR-based, and carbon-based biosensors. Appl Microbiol Biotechnol. 2022;106(8):2827–53.
18. Kumari S, Mishra A. Heavy metal contamination. In: Soil contamination-threats and sustainable solutions. IntechOpen; 2021.
19. Mishra RK, Mentha SS, Misra Y, Dwivedi N. Emerging pollutants of severe environmental concern in water and wastewater: a comprehensive review on current developments and future research. Water-Energy Nexus; 2023.
20. Narang G, Bansal D, Joarder S, Singh P, Kumar L, Mishra V, Singh S, Tumba K, Kumari K. A review on the synthesis, properties, and applications of graphynes. FlatChem. 2023;40:100517.
21. Kumar A, Kataria R. MOFs as versatile scaffolds to explore environmental contaminants based on their luminescence bustle. Sci Total Environ. 2024;172129.
22. Shi W, Li W, Nguyen W, Chen W, Wang J, Chen M. Advances of metal organic frameworks in analytical applications. Mater Today Adv. 2022;15:100273.
23. Li W, Zhu Z, Chen Q, Li J, Tu M. Device fabrication and sensing mechanism in metal-organic framework-based chemical sensors. Cell Rep Phys Sci; 2023.
24. Sohrabi H, Sani PS, Orooji Y, Majidi MR, Yoon Y, Khataee A. MOF-based sensor platforms for rapid detection of pesticides to maintain food quality and safety. Food Chem Toxicol. 2022;165:113176.
25. Karmakar A, Samanta P, Dutta S, Ghosh SK. Fluorescent "turn-on" sensing based on metal–organic frameworks (MOFs). Chem Asian J. 2019;14(24):4506–19.
26. Rebecca PB, Durgalakshmi D, Balakumar S, Rakkesh RA. Advancing healthcare applications: wearable sensors utilizing metal–organic frameworks. Sensors Diagn. 2023;2(6):1360–75.
27. Cheng W, Tang X, Zhang Y, Wu D, Yang W. Applications of metal-organic framework (MOF)-based sensors for food safety: enhancing mechanisms and recent advances. Trends Food Sci Technol. 2021;112:268–82.
28. Gao Y, Zhu Y, Li T, Chen Z, Jiang Q, Zhao Z, et al. Unraveling the high-activity origin of single-atom iron catalysts for organic pollutant oxidation via peroxymonosulfate activation. Environ Sci Technol. 2021;55(12):8318–28.
29. Zhao D, Yu S, Jiang WJ, Cai ZH, Li DL, Liu YL, Chen ZZ. Recent progress in metal-organic framework based fluorescent sensors for hazardous materials detection. Molecules. 2022;27(7):2226.
30. Wu Y, Song Q, Wu J, Zhou J, Zhou L, Wu W. Field study on the soil bacterial associations to combined contamination with heavy metals and organic contaminants. Sci Total Environ. 2021;778:146282.
31. Liu KG, Bigdeli F, Sharifzadeh Z, Gholizadeh S, Morsali A. Role of metal-organic framework composites in removal of inorganic toxic contaminants. J Clean Prod. 2023;404:136709.
32. Zhang X, Zhang S, Tang Y, Huang X, Pang H. Recent advances and challenges of metal–organic framework/graphene-based composites. Compos Part B. 2022;230:109532.

33. Wang B, Lv XL, Feng D, Xie LH, Zhang J, Li M, et al. Highly stable Zr (IV)-based metal–organic frameworks for the detection and removal of antibiotics and organic explosives in water. J Am Chem Soc. 2016;138(19):6204–16.
34. Daniel M, Mathew G, Anpo M, Neppolian B. MOF based electrochemical sensors for the detection of physiologically relevant biomolecules: an overview. Coord Chem Rev. 2022;468:214627.
35. Chabalala MB, Mothudi BM, Ntsendwana B. MOF based nanocomposites for photocatalytic degradation of pollutants in water: a critical review. J Photochem Photobiol A Chem. 2023;115244.
36. Shen Y, Tissot A, Serre C. Recent progress on MOF-based optical sensors for VOC sensing. Chem Sci. 2022;13(47):13978–4007.
37. Zhang S, Wang M, Wang X, Song J, Yang X. Electrocatalysis in MOF films for flexible electrochemical sensing: a comprehensive review. Biosensors. 2024;14(9):420.
38. Liang J, Liang K. Biocatalytic metal–organic frameworks: prospects beyond bioprotective porous matrices. Adv Funct Mater. 2020;30(27):2001648.
39. Majhi SM, Ali A, Rai P, Greish YE, Alzamly A, Surya SG, et al. Metal–organic frameworks for advanced transducer based gas sensors: review and perspectives. Nanosc Adv. 2022;4(3):697–732.
40. Mouhamed AA, Nadim AH, Mahmoud AM, Mostafa NM, Eltanany BM. Bimetallic MOF-based electrochemical sensor for determination of paracetamol in spiked human plasma. BMC Chem. 2024;18(1):148.
41. Wang F, Xiang L, Leung KSY, Elsner M, Zhang Y, Guo Y, et al. Emerging contaminants: a one health perspective. The Innovation; 2024.
42. Baranwal J, Barse B, Gatto G, Broncova G, Kumar A. Electrochemical sensors and their applications: a review. Chemosensors. 2022;10(9):363.
43. Singh R, Gupta R, Bansal D, Bhateria R, Sharma M. A review on recent trends and future developments in electrochemical sensing. ACS Omega. 2024;9(7):7336–56.
44. Zhou Z, Xu T, Zhang X. Empowerment of AI algorithms in biochemical sensors. TrAC Trends Anal Chem. 2024;117613.
45. Li Z, Zhang S, Yu T, Dai Z, Wei Q. Aptamer-based fluorescent sensor array for multiplexed detection of cyanotoxins on a smartphone. Anal Chem. 2019;91(16):10448–57.
46. Pathak AK, Swargiary K, Kongsawang N, Jitpratak P, Ajchareeyasoontorn N, Udomkittivorakul J, Viphavakit C. Recent advances in sensing materials targeting clinical volatile organic compound (VOC) biomarkers: a review. Biosensors. 2023;13(1):114.
47. Malik S, Singh J, Goyat R, Saharan Y, Chaudhry V, Umar A, et al. Nanomaterials-based biosensor and their applications: a review. Heliyon; 2023.
48. Li M, Dong C, Yang J, Yang T, Bai F, Ning Z, et al. Solvothermal synthesis of La-based metal-organic frameworks and their color-tunable photoluminescence properties. J Mater Sci Mater Electron. 2021;32:9903–11.

Advantages and Disadvantages of Metal Organic Framework [MOFs]

13

Abstract

Metal-organic frameworks (MOFs) have emerged as promising materials in various scientific fields due to their porous structures and tenable properties. In forensic science, MOFs have shown potential in applications such as drug delivery, sensing, and storage of volatile compounds. This abstract outlines the advantages and disadvantages of MOFs in forensic science. The advantages encompass their high surface area, tailored functionality, and selectivity in capturing and detecting specific compounds critical in forensic investigations. However, challenges including stability under varying conditions, scalability, and synthesis complexity pose limitations to their widespread implementation. This review highlights the current landscape of MOFs in forensic science, emphasizing their potential and the areas that necessitate further research and development for practical utilization.

13.1 Introduction

Metal organic frameworks are referred to as "MOFs"; this term is usually used to refer to the class of compounds. If an ordinal number comes after it, it denotes a particular metal organic framework. Analyzing the properties and configurations of MOFs can help create framework structures with the desired qualities, like the isoreticular metal–organic frameworks (MOFs) in the IRMOF-1–IRMOF-16 family that have the same symmetry [1–4]. Depending on where they were found, a number of MOFs are assigned to organizations (materials of the Lavoisier Institute), such as UiO, MIL, HKUST, and LIC. Researchers from China and Russia commonly refer to "metalorganic coordination polymers" with specific compositions. Zeolite topology [5] is the other large family of MOFs. MOFs are crystalline materials with a porous structure made of metal ions or clusters linked by organic ligands

V. Mishra, A. Mishra, *Metal-Organic Frameworks as Forensic Detectors*,
https://doi.org/10.1007/978-981-96-2147-7_13

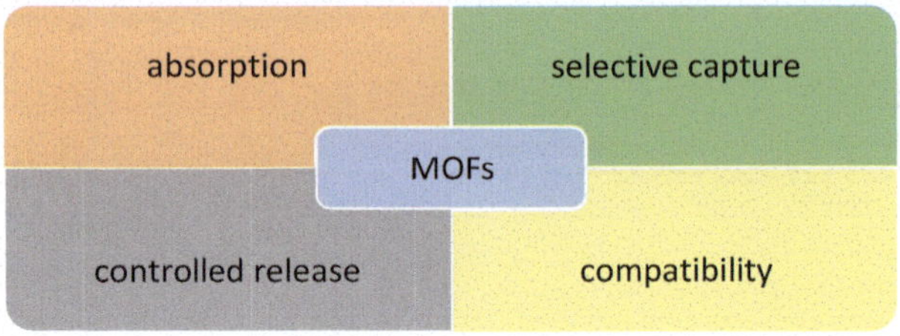

Fig. 13.1 Properties of MOFs in field of forensic sciences

[1–4]. They resemble scaffolds with regularly spaced pores that can be tailored for various applications like gas storage, separation [6–8], catalysis [9–12] and drug delivery. Picture a mesh-like structure with repeating units forming a network of interconnected spaces. MOFs (Metal-Organic Frameworks) are crucial in forensic science due to their exceptional properties. Due to their physicochemical and optical properties, they have been recently proposed for finger-mark detection [13–17]. Their high surface area, tuneable structures, and ability to capture specific molecules make them valuable for:

Adsorption: MOFs can trap and store volatile compounds, aiding in the analysis of trace evidence like explosives, drugs, or residues at crime scenes (Fig. 13.1).

Selective capture: Tailored pore structures enable selective adsorption of specific molecules, enhancing the identification of target compounds amidst complex matrices.

Controlled release: They allow controlled release of stored compounds, aiding in controlled drug delivery or releasing trapped chemicals for further analysis.

Compatibility: MOFs work well with various detection techniques, such as chromatography or spectroscopy, improving the accuracy and sensitivity of forensic analysis.

In essence, MOFs offer precise and versatile tools that help forensic scientists in identifying, isolating, and studying trace substances critical in criminal investigations.

MOFs play a significant role in various branches of forensic science, but they are particularly important in forensic chemistry and trace evidence analysis. In these fields, MOFs are utilized for their ability to capture, separate, and analyze trace substances present at crime scenes. They aid in identifying and characterizing compounds like explosives, drugs, toxins, or residues, thus contributing to the investigation and resolution of criminal cases (Fig. 13.2).

13.2 Applications of MOFs

Metal-organic frameworks (MOFs) have diverse applications in chemistry:

Gas Storage and Separation: MOFs can store gases like hydrogen, methane, or carbon dioxide, aiding in gas purification and separation processes due to their porous structures (Fig. 13.3).

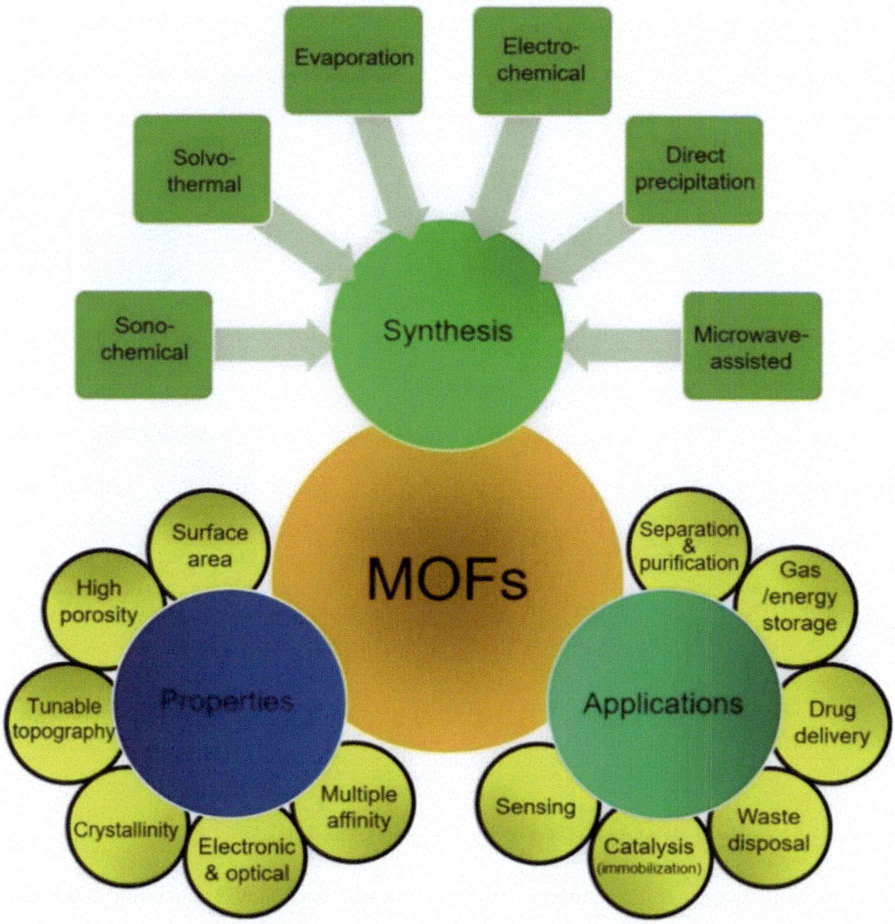

Fig. 13.2 Synthesis, application, properties of MOFs in forensic sciences

Catalysis: They serve as catalysts in various reactions due to their high surface area and tenable pore sizes, enhancing reaction rates and selectivity [18].

Drug Delivery: MOFs offer a platform for controlled drug release by encapsulating pharmaceuticals within their structures, ensuring targeted delivery.

Sensing and Detection: MOFs can be tailored to detect specific molecules or ions, making them useful in sensing applications like environmental monitoring 19 or medical diagnostics 20–22.

Chemical Sensing and Capture: They can selectively capture and separate specific chemicals or pollutants from mixtures, aiding in purification processes 23.

Electrocatalysis: MOFs are employed in electrochemical reactions, particularly in energy-related technologies like batteries, fuel cells, and supercapacitors 24.

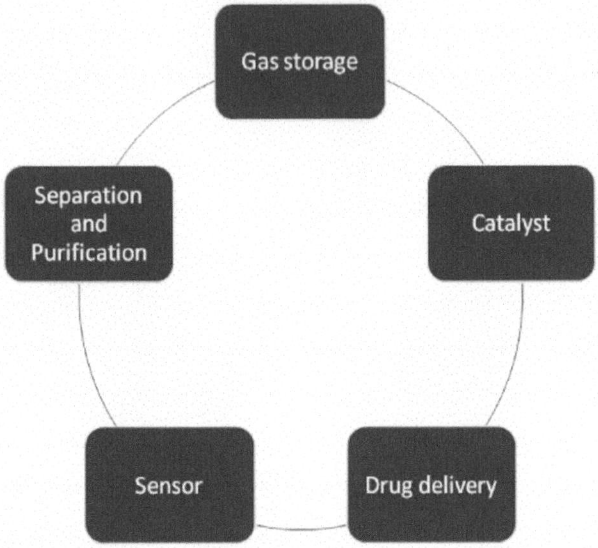

Fig. 13.3 Applications of MOFs in forensic sciences

Cryogenics: MOFs can be used in low-temperature applications, such as in cryogenic storage due to their ability to adsorb gases even at extremely low temperatures 25.

These applications highlight the versatility and potential impact of MOFs across various fields within chemistry.

13.2.1 Samples and Substrates with Fingerprints

Eight donors were used to gather natural, charged, and protein-enriched markers. The former were taken from donors who had not cleansed their hands in the preceding hour, touched their faces on purpose, or worn gloves. Prior to deposition, the donor rubbed their fingers across their forehead to prepare charged finger marks. The mark was to be placed on the substrate by the donors lightly putting their fingers on it [71].

13.2.2 Optical Characterization of MOFs

The UV region of the spectrum has a broad excitation band from approximately 250 to 340 nm with a maximum at 310 nm, and the emission is characteristic of terbium complexes with several narrow peaks from 485 to 635 nm, with the most intense emission peak is between 485 and 635 nm. 547 nm.

Results were observed using available forensic light sources. UV light was generated using a Polylight PL550XL's 350 nm filter.

13.2.3 Mechanism of Interaction

By stamping pure protein solutions like BSA encouraged the production of MOFs on a variety of substrates in their initial paper. It was shown that the presence of proteins chemically induced MOF development. It is not as simple to translate this procedure to finger mark residue, though. Although finger mark secretions contain proteins and amino acids, these substances represent just trace amounts of a complex mixture of numerous other substances.

Using the fluorescence quenching effect, a potent red-fluorescent europium-organic framework (Eu-MOF) material is exploited for the sensitive and selective detection of picric acid, TNT, and tetryl. According to the results, Eu-MOF can be employed with low detection limits (20–140 μg/mL) and good quenching efficiency (Ksv > 104 M − 1) for the visual detection of these three explosives. Moreover, recycling this material only requires rinsing it in ethanol. Additionally covered is the mechanism of fluorescence quenching. Crucially, we built a portable test strip that can detect TNT, and by employing it, we are able to quickly test for explosives on the field. Eu-MOF is therefore a material that shows promise for application as an explosive optical sensor.

Of the technologies in development, a new type of porous hybrid materials called Metal-Organic Frameworks (MOFs) is demonstrating tremendous promise for the removal of pollutants in tertiary treatment procedures. Widespread research has been done on MOFs in a variety of applications, including gas sorption, energy storage, medication delivery and storage, photocatalysis, sensing, and separation. Because of its structure and simplicity in functionalization, MOFs have a wide range of uses. In order to generate an iso reticular structure, MOFs are primarily made up of inorganic metal clusters that are joined by organic linkers through coordination bonds. The kind of shape that is created in the coordination material is what makes the inorganic cluster relevant for this material. The material may readily alter its chemical characteristics inside the framework of the material itself while retaining high porosity and surface area thanks to this connection, which is known as an organic ligand [4].

13.2.4 UiO-Zr MOFs in Water Purification: Applications and Features

The UiO family of MOFs, namely the UiO-66, UiO-67, and derived materials, have been tested for the sorption of various pollutants, including organic pollutants like drugs, organophosphorus compounds, dyes, and hydrocarbons, as well as heavy metals or ions, because of the previously mentioned properties of these materials. We have gathered data regarding the use of these materials in the sorption of various contaminants. Included are the kind and primary attributes of MOFs as stated by the authors, together with information on their maximal sorption capacity and the pollutant's driving force during sorption [4] (Table 13.1).

Table 13.1 Different pollutants adsorbed on Zr MOFs

Type of pollutant	Pollutant
Organic pollutant	Carbamazepine
	Atrazine
	Tetracycline
	Tetracycline hydrochloride
	Phenol
	4-Nitrophenol
	2,4-Dinitrophenol
	2,4,6-Trinitrophenol
	Benzene
	Toluene
	Ethylbenzene
	Xylenes
	Glyphosate
	Triclosan
Heavy metal	Hg (II)
	Pb (II)
	Cd (II)
	Cr (VI)
	Hg0
Metalloid	As (V)
	As (III)
	Sb (III)
Ion	Fluoride
	Phosphate
Dye	Methylene blue
	Methyl orange
	Rhodamine B
Others	p-Arsanilic acid
	Diclofenac sodium
	Methyl-chloro-phenoxy-propionic acid MCPP)

13.3 MOFs: A Versatile Platform for Forensic Detections

Numerous forensic laboratories worldwide frequently employ several current techniques that have been adapted from other fields. Similarly, a recent material sciences paper addressed the guided growth of Metalorganic Frameworks (MOFs). The crystal structure of MOFs is highly porous, and they have a large interior surface area. They are especially useful in many different forensic industry applications because of these attributes [70].

Several techniques can identify MOFs, including:

X-ray Diffraction (XRD): Determines the crystal structure and provides information about the arrangement of atoms within the MOF.

Scanning Electron Microscopy (SEM): Provides surface morphology and size information of the MOF crystals.

Nuclear Magnetic Resonance (NMR) Spectroscopy: Analyzes the structure and composition of the MOFs, providing information about chemical environments and connectivity.

Fourier Transform Infrared (FTIR) Spectroscopy: Detects functional groups present in MOFs, aiding in their identification.

Thermogravimetric Analysis (TGA): Determines the thermal stability and decomposition temperature of MOFs.

Gas chromatography-mass spectrometry (GC-MS): It can be used to identify Metal-Organic Frameworks (MOFs) by analyzing their volatile components or byproducts. However, it's crucial to note that MOFs are typically non-volatile and might not be directly amenable to GC-MS analysis. Instead, if there are volatile compounds associated with the synthesis, degradation, or modification of MOFs, GC-MS could potentially identify those compounds. The identification of MOFs themselves might require other analytical techniques like X-ray diffraction (XRD) or infrared spectroscopy (IR).

Each of these techniques offers distinct information about MOFs, aiding in their identification and characterization.

MOFs, while promising, come with drawbacks in forensic techniques. Their limited stability under varying environmental conditions like humidity or pH levels could affect their performance and reliability. Also, MOFs might not always exhibit the necessary selectivity for specific compounds, which could hinder their accuracy in identifying target substances in complex forensic samples. Additionally, their synthesis can be complex and costly, potentially limiting their widespread application in forensic analysis.

While X-ray diffraction (XRD) is a powerful technique for characterizing MOFs, it does have limitations when applied to these materials:

Amorphous or Poorly Crystalline MOFs: Some MOFs might exhibit poor crystallinity, making it challenging to obtain high-quality XRD patterns or fully resolve their structures.

Large Unit Cells: MOFs with large unit cells can pose challenges in obtaining accurate structural information due to limitations in resolution or data collection.

Sample Preparation: Preparing MOF samples for XRD analysis can be complex and time-consuming, potentially requiring special techniques like solvent removal or encapsulation.

Low Signal-to-Noise Ratio: Obtaining good quality diffraction patterns for MOFs might be hindered by low signal-to-noise ratios, affecting the accuracy of structural determination.

Despite these limitations, XRD remains a valuable tool for understanding the structure and crystalline nature of MOFs. Advances in techniques and instrumentation continue to address some of these challenges.

In Fourier Transform Infrared Spectroscopy (FTIR), MOFs might present certain limitations:

Low Sensitivity: Some MOFs could exhibit weak absorption bands in the infrared region, making it challenging to detect specific functional groups or subtle structural changes.

Overlap of Bands: Complex MOF structures might result in overlapping absorption bands in the FTIR spectra, making it difficult to accurately identify individual functional groups or components.

Sample Preparation: The preparation of MOF samples for FTIR analysis might require careful handling and preparation techniques to ensure representative spectra, especially if the MOF is sensitive to environmental factors or requires specific handling conditions.

Despite these limitations, FTIR is still valuable for identifying functional groups and providing insights into the chemical composition of MOFs, albeit with potential challenges in certain cases.

In Nuclear Magnetic Resonance (NMR) spectroscopy, MOFs might have limitations:

Signal Broadening: Some MOFs, especially those with paramagnetic metal centers or large molecular sizes, might exhibit broadened or overlapping NMR signals, making it challenging to resolve individual peaks and interpret spectra accurately.

Pore Blocking: In certain cases, guest molecules or solvents within the pores of MOFs might hinder NMR analysis by blocking access to certain sites or causing line broadening due to restricted molecular motion.

Signal Weakness: Some MOFs might yield weak NMR signals, requiring longer acquisition times or specialized NMR techniques to enhance signal intensity and improve spectral resolution.

Dynamic Instability: Some MOFs might display dynamic structural changes or flexibility, resulting in broad or shifting signals in the NMR spectra, complicating the interpretation of their structures.

Despite these challenges, NMR spectroscopy remains a valuable technique for characterizing MOFs, providing insights into their structures, connectivity, and chemical environments. Advanced NMR methodologies and techniques continue to address some of these limitations.

In Scanning Electron Microscopy (SEM), MOFs might pose certain limitations:

Sample Conductivity: MOF samples can be insulating, leading to charging effects during SEM analysis, which might cause image distortion or artifacts.

Sample Preparation: Preparing MOF samples for SEM analysis often involves coating them with a conductive layer (like gold or carbon) to mitigate charging effects, but this process can potentially alter the surface morphology or introduce artifacts.

Resolution Limits: Resolving fine details or structures within MOFs might be limited due to the inherent resolution constraints of SEM, which could affect the visualization of small features or pore structures within the MOF.

Surface Sensitivity: SEM provides surface morphology information but might not provide detailed insights into the internal structure or composition of MOFs.

Despite these limitations, SEM remains a valuable tool for observing the surface morphology and general structure of MOFs, allowing for qualitative characterization, and understanding of their external features.

Using Metal-Organic Frameworks (MOFs) in Gas Chromatography-Mass Spectrometry (GC-MS) faces several challenges and disadvantages:

Non-volatility: MOFs are typically non-volatile, making it challenging to directly analyze them using GC-MS as the components may not readily vaporize under the GC-MS conditions.

Thermal Stability: Some MOFs might not withstand the high temperatures required for GC-MS analysis, leading to decomposition or alteration of their structure during the analysis process.

Complexity: MOFs can be structurally complex, containing multiple components, which might not be easily separable or detectable using GC-MS due to their similarity in mass or fragmentation patterns.

Detection Sensitivity: The low concentration or trace amounts of volatile components or byproducts associated with MOFs might make their detection and analysis challenging using GC-MS, affecting sensitivity and accuracy.

Sample Preparation: Extracting volatile components or byproducts from MOFs for GC-MS analysis might require complex sample preparation steps, potentially introducing variability or loss of analytes during the process.

These limitations collectively pose challenges when attempting to directly analyze MOFs using GC-MS, necessitating complementary or alternative analytical techniques for comprehensive characterization.

13.4 Photochemical and Electrochemical Properties of MOFs

Metal organic frameworks (MOFs) are a class of porous crystalline materials with well-organized structures and unique optical, electrical, and catalytic capabilities. Both electrochemical and photoelectrochemical (PEC) pathways have demonstrated the high effectiveness of MOFs and materials derived from MOFs as catalysts for water splitting. To maximize their performance, these MOFs can also have their photochemical and electrochemical capabilities optimized by modifying their organic ligands and/or metal centers, bandgap, surface area, current density, electrochemical active surface area, and overpotential [70]. Even though this field is still in its infancy, there is a growing amount of interest in this subject because of promising results that indicate significant future applications.

13.4.1 Advantages

Metal-organic frameworks (MOFs) offer promising advantages in forensic science due to their unique photochemical and electrochemical properties. Photochemical properties enable MOFs to undergo light-induced reactions, allowing for applications in forensic analysis, such as photo-induced degradation of pollutants or luminescent tagging for tracking and identifying substances.

On the other hand, MOFs' electrochemical properties facilitate their use in sensors for detecting trace amounts of compounds, aiding in forensic investigations by providing sensitive and selective detection methods for analyzing evidence.

These properties empower MOFs to potentially revolutionize forensic analysis by offering advanced tools for identification, detection, and analysis of substances crucial in solving crimes or analyzing evidence.

While metal-organic frameworks (MOFs) exhibit promising photochemical and electrochemical properties in forensic science, there are some potential drawbacks to consider:

Stability: Some MOFs might lack long-term stability, especially in harsh environmental conditions or when exposed to certain chemicals, which could affect their reliability in forensic applications over time.

Synthesis Challenges: The synthesis of MOFs can be complex and may require specific conditions, expertise, and time, which could limit their widespread adoption in forensic labs due to practical challenges in production.

Cost: Some MOFs might involve expensive materials or intricate fabrication processes, making them costly for routine forensic analyses and limiting their accessibility in budget-constrained environments.

Sensitivity to Contaminants: MOFs might be sensitive to certain contaminants or impurities present in forensic samples, leading to interference or inaccurate readings during analysis.

Addressing these challenges through advancements in MOF synthesis techniques, stability improvements, and cost-effective production methods could enhance their viability and efficacy in forensic science applications.

13.4.2 MOFs for Detection of Latent Fingerprint Detection

Currently used as sensors, separators in membranes, and catalysts, MOFs are crystalline porous structures. Recent proposals for fingerprint detection have been based on their optical and physicochemical properties. In this work, their potential for fingerprint detection was examined in more detail. Both naturally occurring fingerprints and marks that were charged and protein-enriched were used to assess the technique's efficacy. A number of parameters, such as pH, immersion time, precursor concentration, and detection methods, were investigated and improved. The performance of the enhanced MOF-based approach was then compared to that of currently popular approaches. The findings demonstrated the high specificity with which MOFs can identify fingerprints, especially those that are rich in proteins and are deposited by biological fluids [71]. However, results were discovered to be less than those from the detection techniques currently in use after.

However, after comparison and evaluation against benchmark approaches, results were found to be lower than those from currently used detection methods. With further research and development, MOFs, however, may prove to be a competitive alternative to the current powder suspension techniques. Recently, a highly luminous red emissive nanocomposite material was created by successfully combining carbonized PPDCD polymer dots from the Zhu group in situ with the zeolitic imidazolate framework ZIF-8. The resulting material shows a strong and consistent red emission at solid state after conjugation with ZIF-8, with the largest peak centered at 630 nm. Furthermore, PPDCDs exhibit a significantly higher fluorescence quantum yield, increasing from 17.8% to 43.5%. This material's extremely high sensitivity has been successfully applied to the detection of LFPs on marble [71, 72].

13.5 Metal-Organic Frameworks for Detection of Pesticides in Forensic Samples

Anthropogenic processes must be accelerated due to the growing global population, which pollutes terrestrial and marine environments with dangerous organic compounds. Particularly dangerous man-made chemicals are called persistent organic pollutants (POPs). POP use is currently strictly regulated or outright prohibited worldwide due to its relatively high chemical resistance. Sadly, certain POPs are still employed as practise aids in agriculture. POPs were primarily used as pesticides in the past. As a result, the need for environmental monitoring is rising in order to precisely identify and measure pesticidal POPs. Target molecules should be able to be detected through host-guest interactions down to the nanomolar range thanks to the selectivity and sensitivity of chemical sensors and adsorption materials made for specific pesticides. This could be accomplished by materials with a very large active surface area, a well-defined structure, and exceptional stability. Within that framework, novel materials under investigation are referred to as metal-organic frameworks (MOFs). It is possible to functionalize the structures of various MOFs to facilitate desired host-guest interactions. In this book's Chap. 6, we critically evaluate the application of MOFs for the detection and adsorption of particular pesticides that the Stockholm Convention classifies as persistent organic pollutants (POPs). Regarding this, they employed the Au-Ag/Si/zeolitic imidazolate framework-8 (ZIF-8) assembly (refer to Fig. 13.4) as a SERS material in their PCP detection method. This material was prepared layer by layer with a great deal of control over the film thickness and, thus, the SERS effect utilized. It was found that ZIF-8 and PCP's electrostatic interactions led to a concentration of the target molecule around Au-Ag nanoparticles, plasmonic effects that greatly enhanced Raman scattering. Application of ZIF-8/Au-Ag/Si assembly for sub-pM PCP detection using SERS [73].

13.5.1 Advantages

MOFs present several advantages for detecting pesticides in forensic samples. Their high surface area and customizable structures enable selective adsorption of specific

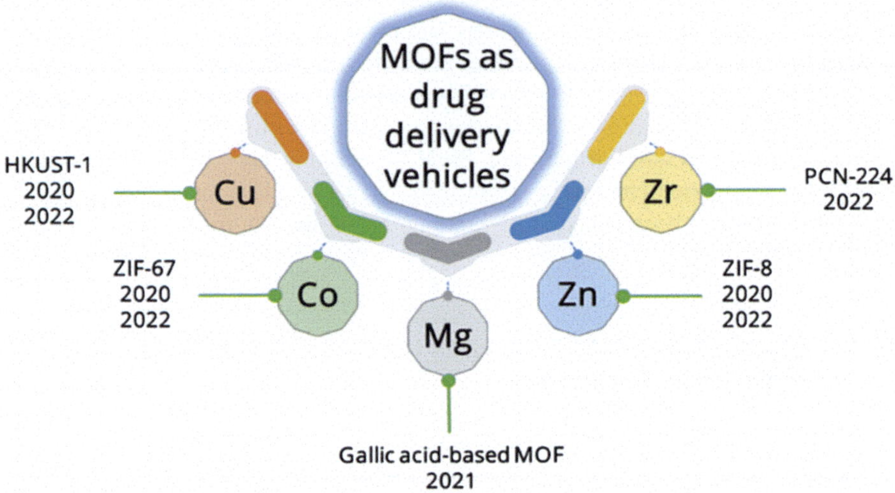

Fig. 13.4 MOFs as drug delivery vehicles

pesticides, aiding in their extraction and concentration from complex matrices. MOFs can be tailored to target certain pesticides, enhancing their sensitivity and accuracy in detection. Additionally, their porosity facilitates the controlled release of trapped compounds, enabling efficient extraction and analysis of pesticides from forensic samples [73].

13.5.2 Disadvantages

Some limitations exist when using MOFs for pesticide detection in forensic samples. MOFs can be sensitive to environmental conditions, affecting their stability and performance during the analysis of complex matrices. Their synthesis can also be intricate and time-consuming, potentially impeding their widespread adoption for routine forensic analysis. Moreover, the selectivity of certain MOFs might be limited to specific pesticides, requiring the development of multiple MOF types for comprehensive pesticide detection, which could be resource-intensive [73].

13.6 Metal-Organic Frameworks for Detection of Antibiotics in Forensic Samples

In recent years, there has been a great deal of research interest in the creation of suitable chemical sensors for the selective detection and quantification of antibiotics due to the serious environmental threat that antibiotics pose. The overabundance of antibiotics in the soil, brought on by increased human use and animal activity, may have detrimental effects on living systems, which is why appropriate sensors must be developed. The many detection technologies currently in use for the identification of antibiotics have been joined by the fluorescence technique as a practical and effective substitute. Recently, a large variety of fluorescence sensors have been developed for the specific detection of antibiotics in solution and vapour phase. By employing luminous metal-organic frameworks (LMOFs) as fluorescence sensors for the specific identification of different types of antibiotics, we have achieved noteworthy. The LMOFs-based fluorescence sensors discussed are chosen based on their ability to detect a particular class of antibiotics [74]. The essay also addresses the field's future directions and provides an overview of the advantages and present challenges of using fluorescence sensors based on light metal oxide fragments (LMOFs) for the detection of antibiotics on surfaces such as glass. The obtained fluorescent images merely illustrated the complex characteristics of LFPs with levels 1 through 3. PPDCDs@ZIF-8 offers a fast and accurate way to view the sweat pores in LFPs, which makes it a valuable tool for crime scenes. The MOF-76 was employed as a colorimetric indicator to distinguish between various antibiotic species in an aqueous solution (Fig. 13.5). The spin-orbit coupling that the BTC ligand promoted allowed for the intersystem crossing of the n-p* transition (S1-T1). This luminophore could repeatedly detect the antibiotics ciprofloxacin (CPFX), nitroimidazole (NDZ), and nitrofurans (NFs) by employing three different detection activities. Bathochromic shift 14 of the ligand is used by MOF-76 to identify NFs, while color changes in CPFX's luminescence and lanthanide luminescent quenching are used to detect NDZ. The sensitivity and recyclable nature of the MOF76 were also examined, and it was found that the antibiotics it detected had a detection limit of ppb [74, 75].

13.6.1 Advantages

MOFs offer several advantages for detecting antibiotics in forensic samples. Their high surface area and tunable structures enable selective adsorption of specific antibiotics, aiding in their extraction and concentration from complex matrices. MOFs can be customized to target particular antibiotics, enhancing sensitivity and accuracy in detection. Moreover, their porosity facilitates controlled release, allowing efficient extraction and analysis of antibiotics from forensic samples. This versatility makes MOFs promising for precise and tailored antibiotic detection in forensic investigations [76].

13.6.2 Disadvantages

When applied to detect antibiotics in forensic samples, MOFs face several challenges. One drawback is their selectivity, as certain MOFs might not be tailored to detect a wide range of antibiotics, necessitating the development of various MOF structures for comprehensive detection. Additionally, MOFs' susceptibility to environmental conditions like humidity and pH variations might impact their stability

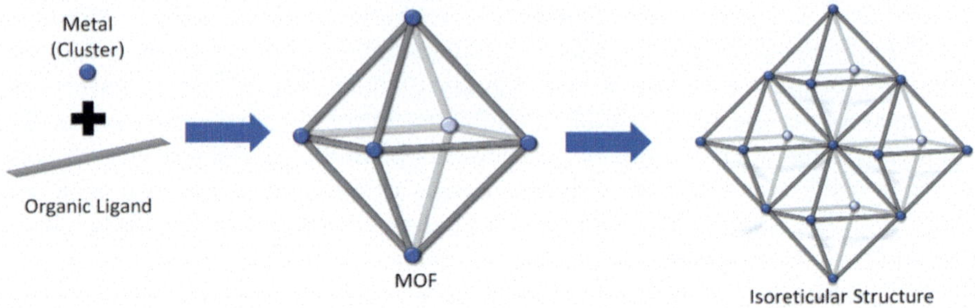

Fig. 13.5 Scheme of MOF, formed by organic ligands and metallic clusters combining into an isoreticular structure

Table 13.2 Advantages and disadvantages of MOFs

Advantages of MOFs	Disadvantages of MOFs
1. High surface area for adsorption	1. Sensitivity to environmental conditions
2. Tailored pore structures for specific analytes	2. Stability concerns in varying pH or humidity
3. Tuneable properties for selectivity	3. Synthesis complexity and cost
4. Potential for controlled release of trapped compounds	4. Limited long-term stability
5. Compatibility with various detection techniques	5. Challenges in scaling production

and performance during the analysis of complex forensic samples containing antibiotics. Synthesizing MOFs with specific affinities for different antibiotics can also be laborious and time-consuming, potentially hindering their practical use in routine forensic analysis [76] (Table 13.2).

13.7 Conclusion

Metal-organic frameworks (MOFs) offer several advantages in forensic science. They possess high surface areas for efficient adsorption of analytes, aiding in sample collection and extraction. MOFs can selectively capture specific compounds, aiding in separation and identification. However, limitations include stability issues in harsh conditions and potential interference with complex samples. Overall, while MOFs show promise in forensic applications, further research is needed to address their drawbacks and maximize their potential.

References

1. Hu C, Xiao J-D, Mao X-D, Song L-L, Yang X-Y, Liu S-J. Toughening mechanisms of epoxy resin using aminated metal-organic framework as additive. Mater Lett. 2019;240:113–6.
2. Xing X-S, Fu Z-H, Zhang N-N, Yu X-Q, Wang M-S, Guo G-C. High proton conduction in an excellent water-stable gadolinium metal–organic framework. Chem Commun. 2019;55(9):1241–4.

3. Mirkovic I, Lei L, Ljubic D, Zhu S. Crystal growth of metal–organic framework5 around cellulose-based fibers having a necklace morphology. ACS Omega. 2019;4(1):169–75.
4. Yang H, et al. Metal–organic framework coated titanium dioxide nanorod array p–n heterojunction photoanode for solar water-splitting. Nano Res. 2019;12:643–50.
5. Chatterjee S, Jeevanandham S, Mukherjee M, Vo DVN, Mishra V. Significance of re-engineered zeolites in climate mitigation–A review for carbon capture and separation. J Environ Chem Eng. 2021; 9(5):105957.
6. Kampouri S, et al. Dual-functional photocatalysis: concurrent photocatalytic hydrogen generation and dye degradation using MIL-125-NH2 under visible light 19Irradiation. Adv Funct Mater. 2018;28(52):1870373.
7. Chao M-Y, et al. A single-crystal to single-crystal conversion scheme for two-dimensional metal–organic framework bearing linear Cd3 secondary building units. Cryst Growth Des. 2018;19(2):724–9.
8. Huan W, Xing M, Cheng C, Li J. Facile fabrication of magnetic metal–organic framework nanofibers for specific capture of phosphorylated peptides. ACS Sustain Chem Eng. 2018;7(2):2245–54.
9. Yang Q, Yang C, Lin C, Jiang H. Metal–organic-framework-derived hollow N-doped porous carbon with ultrahigh concentrations of single Zn atoms for efficient carbon dioxide conversion. Angew Chem. 2019;131(11):3549–53.
10. Rajak R, Saraf M, Mobin SM. Robust heterostructures of a bimetallic sodium–zinc metal–organic framework and reduced graphene oxide for high-performance supercapacitors. J Mater Chem A. 2019;7(4):1725–36.
11. Yang F, Wu M, Wang Y, Ashtiani S, Jiang H. A GO-induced assembly strategy to repair MOF nanosheet-based membrane for efficient H2/CO2 separation. ACS Appl Mater Interfaces. 2018;11(1):990–7.
12. Dwivedi, B., Mishra, P. K., & Mishra, V. (2022). Chiral metal-organic frameworks for asymmetrical catalysis. In Metal-Organic Frameworks (MOFs) as Catalysts (pp. 493-513). Singapore: Springer Nature Singapore.
13. Yadav N, Mudgal D, Mishra A, Shukla S, Malik T, Mishra V. Harnessing fluorescent carbon quantum dots from natural resource for advancing sweat latent fingerprint recognition with machine learning algorithms for enhanced human identification. Plos one, 2024;19(1):e0296270.
14. Li L, Zhang S, Ruffley JP, Johnson JK. Energy efficient formaldehyde synthesis by direct hydrogenation of carbon monoxide in functionalized metal–organic frameworks. ACS Sustain Chem Eng. 2018;7(2):2508–15.
15. Wang C-Y, Yu B, Fu H, Wang P, Wang C-C. A mixed valence Tb (III)/Tb (IV) metal–organic framework: crystal structure, luminescence property and selective detection of naproxen. Polyhedron. 2019;159:298–307.
16. Wang Q, Liu DJ, Cui LL, Hu X-L, Wang X-L, Su Z-M. A 3D pillared-layer metal–organic framework with fluorescence property for detection of nitroaromatic explosives. New J Chem. 2019;43(2):963–9.
17. Mahmoodi NM, Taghizadeh M, Taghizadeh A. Activated carbon/metal-organic framework composite as a bio-based novel green adsorbent: preparation and mathematical pollutant removal modeling. J Mol Liq. 2019;277:310–22.
18. Mohaghegh N, Faraji M, Abedini A. Highly efficient multifunctional Ag/TiO 2 nanotubes/Ti plate coated with MIL-88B (Fe) as a photocatalyst, adsorbent, and disinfectant in water treatment. Appl Phys A Mater Sci Process. 2019;125:1–10.
19. Kim M-K, Kim SH, Park M, Ryu SG, Jung H. Degradation of chemical warfare agents over cotton fabric functionalized with UiO-66-NH 2. RSC Adv. 2018;8(72):41633–8.
20. Shen J, Wang N, Wang YG, Yu D, Ouyang X. Efficient adsorption of Pb (II) 20 from aqueous solutions by metal organic framework (Zn-BDC) coated magnetic montmorillonite. Polymers (Basel). 2018;10(12):1383.
21. Esmaeilzadeh M. A composite prepared from a metal-organic framework of type MIL101 (Fe) and morin-modified magnetite nanoparticles for extraction and speciation of vanadium (IV) and vanadium (V). Microchim Acta. 2019;186:1–7.

22. Zhang L, et al. A non-enzymatic voltammetric xanthine sensor based on the use of platinum nanoparticles loaded with a metal-organic framework of type MIL-101 (Cr). Application to simultaneous detection of dopamine, uric acid, xanthine and hypoxanthine. Microchim Acta. 2019;186:1–10.
23. Henrique A, Rodrigues AE, Silva JAC. Separation of hexane isomers in ZIF8 by fixed bed adsorption. Ind Eng Chem Res. 2018;58(1):378–94.
24. Yaghi OM, O'Keeffe M, Ockwig NW, Chae HK, Eddaoudi M, Kim J. Reticular synthesis and the design of new materials. Nature. 2003;423(6941):705–14.
25. Dutta A, Pan Y, Liu J-Q, Kumar A. Multicomponent isoreticular metal-organic frameworks: principles, current status and challenges. Coord Chem Rev. 2021;445:214074.
26. Hu J-S, Lei Z, Pan C-L, He J. A new 2D→ 3D polythreaded framework constructed on an N-centered tripodal linker and copper (i). Mendeleev Commun. 2014;24(5):290–2.
27. Hu J-S, Zhang X-M, Xing H-L, He J, Shi J-J. Construction of an unusual 3D framework based on V-shaped imidazolyl and oxalate ligands. Mendeleev Commun. 2013;4(23):231–2.
28. Yutkin MP, Zavakhina MS, Samsonenko DG, Dybtsev DN, Fedin VP. Porous homo-and heterochiral cobalt (II) aspartates with high thermal stability of the metal-organic framework. Russ Chem Bull. 2010;59:733–40.
29. Zavakhina MS, Samsonenko DG, Dybtsev DN, Yutkin MP, Virovets AV, Fedin VP. Synthesis and structure of chiral coordination polymers of co II, cu II, and mg II saccharates. Russ Chem Bull. 2013;62:716–21.
30. Zavakhina MS, Samsonenko DG, Virovets AV, Dybtsev DN, Fedin VP. Homochiral cu (II) and ni (II) malates with tunable structural features. J Solid State Chem. 2014;210(1):125–9.
31. Chen X-Y, et al. Microporous metal– organic frameworks built on a Ln3 cluster as a six-connecting node. Chem Mater. 2005;17(11):2866–74.
32. Baniani A, Rivera MP, Lively RP, Vasenkov S. Quantifying diffusion of organic liquids in a MOF component of MOF/polymer mixed-matrix membranes by high field NMR. J Memb Sci. 2021;640:119786.
33. Wu J-Y, Chao T-C, Zhong M-S. Influence of counteranions on the structural modulation of silver–di (3-pyridylmethyl) amine coordination polymers. Cryst Growth Des. 2013;13(7):2953–64.
34. Zavyalova AG, Kladko DV, Chernyshov IY, Vinogradov VV. Large MOFs: synthesis strategies and applications where size matters. J Mater Chem A. 2021;9(2145):25258–71.
35. Ivanova II, Kolyagin YG. Application of multinuclear MAS NMR for the in situ monitoring of hydrothermal synthesis of zeolites. Chem Eur J. 2021;27(57):14143–67.
36. Chen W, Du L, Wu C. Hydrothermal synthesis of MOFs. In: Metal-organic frameworks for biomedical applications. Elsevier; 2020. p. 141–57.
37. Hu Z, Kundu T, Wang Y, Sun Y, Zeng K, Zhao D. Modulated hydrothermal synthesis of highly stable MOF-808 (Hf) for methane storage. ACS Sustain Chem Eng. 2020;8(46):17042–53.
38. Zhang Y, Bo X, Nsabimana A, Han C, Li M, Guo L. Electrocatalytically active cobalt-based metal–organic framework with incorporated macroporous carbon composite for electrochemical applications. J Mater Chem A. 2015;3(2):732–8.
39. Varsha MV, Nageswaran G. Direct electrochemical synthesis of metal organic frameworks. J Electrochem Soc. 2020;167(15):155527.
40. Sun P-P, Zhang Y-H, Shi H, Shi F-N. Controllable one step electrochemical synthesis of PANI encapsulating 3d-4f bimetal MOFs heterostructures as electrode materials for high-performance supercapacitors. Chem Eng J. 2022;427:130836.
41. Campagnol N, Souza ER, De Vos DE, Binnemans K, Fransaer J. Luminescent terbium-containing metal–organic framework films: new approaches for theelectrochemical synthesis and application as detectors for explosives. Chem Commun. 2014;50(83):12545–7.
42. MacGillivray LR. Metal-organic frameworks: design and application. Wiley; 2010.
43. Pichon A, Lazuen-Garay A, James SL. Solvent-free synthesis of a microporous metal–organic framework. CrystEngComm. 2006;8(3):211–4.
44. James SL, et al. Mechanochemistry: opportunities for new and cleaner synthesis. Chem Soc Rev. 2012;41(1):413–47.

45. Masoomi MY, Morsali A, Junk PC. Rapid mechanochemical synthesis of two new Cd (II)-based metal–organic frameworks with high removal efficiency of Congo red. CrystEngComm. 2015;17(3):686–92.
46. Phang WJ, Lee WR, Yoo K, Ryu DW, Kim B, Hong CS. pH-dependent proton conducting behavior in a metal–organic framework material. Angew Chem. 2014;126(32):8523–7.
47. Jhung SH, Chang J-S, Hwang JS, Park S-E. Selective formation of SAPO-5 and SAPO-34 molecular sieves with microwave irradiation and hydrothermal heating. Microporous Mesoporous Mater. 2003;64(1–3):33–9.
48. Deeraj BDS, Jayan JS, Raman A, Saritha A, Joseph K. Recent prospects and trends on zeolitic imidazolate frameworks for microwave absorption and EMI shielding applications. Synth Met. 2023;296:117354.
49. Sabouni R, Kazemian H, Rohani S. Microwave synthesis of the CPM-5 metal organic framework. Chem Eng Technol. 2012;35(6):1085–92.
50. Morsali A, Monfared HH, Morsali A, Janiak C. Ultrasonic irradiation assisted syntheses of one-dimensional di (azido)-dipyridylamine Cu (II) coordination polymer nanoparticles. Ultrason Sonochem. 2015;23:208–11.
51. Jung D-W, Yang D-A, Kim J, Kim J, Ahn W-S. Facile synthesis of MOF-177 by a sonochemical method using 1-methyl-2-pyrrolidinone as a solvent. Dalt Trans. 2010;39(11):2883–7.
52. Son W-J, Kim J, Kim J, Ahn W-S. Sonochemical synthesis of MOF-5. Chem Commun. 2008;47:6336–8.
53. Haque E, Khan NA, Park JH, Jhung SH. Synthesis of a metal–organic framework material, iron terephthalate, by ultrasound, microwave, and conventional electric heating: a kinetic study. Chem Eur J. 2010;16(3):1046–52.
54. Pineiro-Lopez L, Arcis-Castillo Z, Muñoz MC, Real JA. Clathration of five-membered aromatic rings in the bimetallic spin crossover metal–organic framework [Fe (TPT) 2/3 {MI (CN) 2} 2]. G (MI= Ag, Au). Cryst Growth Des. 2014;14(12):6311–9.
55. Shi X, Zhu G, Qiu S, Huang K, Yu J, Xu R. Zn2 [(S)-O3PCH2NHC4H7CO2] 2: a homochiral 3d zinc phosphonate with helical channels. Angew Chem. 2004;116(47):6644–7.
56. Jhung SH, Yoon JW, Hwang J-S, Cheetham AK, Chang J-S. Facile synthesis of nanoporous nickel phosphates without organic templates under microwave irradiation. Chem Mater. 2005;17(17):4455–60.
57. Zhang W, Yang Y, Zai S, Weng Ng S, Chen X. Syntheses, structures and magnetic properties of dinuclear copper (II)–lanthanide (III) complexes bridged by 2-Hydroxymethyl-1-methylimidazole. Wiley; 2008.
58. Beamish-Cook J, Shankland K, Murray CA, Vaqueiro P. Insights into the mechanochemical synthesis of MOF-74. Cryst Growth Des. 2021;21(5):3047–55.
59. Głowniak S, Szczęśniak B, Choma J, Jaroniec M. Mechanochemistry: toward green synthesis of metal–organic frameworks. Mater Today. 2021;46:109–24.
60. Kaupp G. Mechanochemistry: the varied applications of mechanical bond-breaking. CrystEngComm. 2009;11(3):388–403.
61. Garay AL, Pichon A, James SL. Solvent-free synthesis of metal complexes. Chem Soc Rev. 2007;36(6):846–55.
62. Willans CE, et al. Tripodal imidazole frameworks: reversible vapour sorption both with and without significant structural changes. Dalt Trans. 2011;40(3):573–82.
63. Kulesa KM, et al. A bioinspired cobalt catalyst based on a tripodal imidazole/pyridine platform capable of water reduction and oxidation. J Inorg Biochem. 2023;242:112162.
64. Berger J, Dönmez A-S, Ullrich A, Bunzen H, Fischer RA, Kieslich G. Reduced thermal expansion by surface-mounted nanoparticles in a pillared-layered metal-organic framework. Commun Chem. 2022;5(1):177.
65. Rohde MD. Identification of structural phases in pillared-layered, Amino-Functionalized M2 (abdc) 2 (dabco)(M= Co, Ni, Cu, Zn). The University of Bergen; 2023.
66. Bonneau M, Lavenn C, Ginet P, Otake K, Kitagawa S. Upscale synthesis of a binary pillared layered MOF for hydrocarbon gas storage and separation. Green Chem. 2020;22(3):718–24.

67. Vaitsis C, Sourkouni G, Argirusis C. Sonochemical synthesis of MOFs. In: Metal-organic frameworks for biomedical applications. Elsevier; 2020. p. 223–44.
68. Yu K, Lee Y-R, Seo JY, Baek K-Y, Chung Y-M, Ahn W-S. Sonochemical synthesis of Zr-based porphyrinic MOF-525 and MOF-545: enhancement in catalytic and adsorption properties. Microporous Mesoporous Mater. 2021;316:110985.
69. Othong J, Boonmak J, Promarak V, Kielar F, Youngme S. Sonochemical synthesis of carbon dots/lanthanoid MOFs hybrids for white light-emitting diodes with high color rendering. ACS Appl Mater Interfaces. 2019;11(47):44421–9.
70. Khan MM, Rahman A, Matussin SN. Recent progress of metal-organic frameworks and metal-organic frameworks-based heterostructures as photocatalysts. Nano. 2022;12(16):2820.
71. Zhu Q, et al. Red emissive nanocomposite with high quantum yield for ultrasensitive and selective detection of latent fingerprints. Microchem J. 2023;190:108688.
72. Kumar A, Sahoo SC, Mehta SK, Soni P, Sharma V, Kataria R. A luminescent Zn-MOF for the detection of explosives and development of fingerprints. Anal Methods. 2022;14(7):700–7.
73. Yan L, Yang P, Cai H, Chen L, Wang Y, Li M. ZIF-8-modified Au–Ag/Si nanoporous pillar array for active capture and ultrasensitive SERS-based detection of pentachlorophenol. Anal Methods. 2020;12(32):4064–71.
74. Marimuthu M, Arumugam SS, Sabarinathan D, Li H, Chen Q. Metal organic framework based fluorescence sensor for detection of antibiotics. Trends Food Sci Technol. 2021;116:1002–28.
75. Yin M, Zhang L, Wei X, Sun J, Xu D. Detection of antibiotics by electrochemical sensors based on metal-organic frameworks and their derived materials. Microchem J. 2022:107946.
76. Yang Y, Zhao L, Sun M, Wei P, Li G, Li Y. Highly sensitive luminescent detection toward polytypic antibiotics by a water-stable and white-light-emitting MOF76 derivative. Dyes Pigments. 2020;180:108444.
77. Islamoglu T, et al. Metal–organic frameworks against toxic chemicals. Chem Rev. 2020;120(16):8130–60.
78. Woellner M, Hausdorf S, Klein N, Mueller P, Smith MW, Kaskel S. Adsorption and detection of hazardous trace gases by metal–organic frameworks. Adv Mater. 2018;30(37):1704679.
79. Khan NA, Hasan Z, Jhung SH. Adsorptive removal of hazardous materials using metal-organic frameworks (MOFs): a review. J Hazard Mater. 2013;244:444–56.
80. Han X, et al. Reversible adsorption of nitrogen dioxide within a robust porous metal–organic framework. Nat Mater. 2018;17(8):691–6.
81. Shi Z-Q, Ji N-N, Hu H-L. Luminescent triphenylamine-based metal–organic frameworks: recent advances in nitroaromatics detection. Dalt Trans. 2020;49(37):12929–39.
82. Sen Bishwas M, Malik M, Poddar P. Raman spectroscopy-based sensitive, fast and reversible vapour phase detection of explosives adsorbed on metal–organic frameworks UiO-67. New J Chem. 2021;45(16):7145–53.
83. Charles S, Geusens N, Nys B. Interpol review of gunshot residue 2019 to 2021. Forensic Sci Int Synergy. 2023;6:100302.
84. de Oliveira Silva JP, Colaço MV, De Souza LA, dos Santos MV, Pugina RS, Marques LF. Exploring the intense lanthanide luminescence and high thermal stability in a new mixed Eu3+/Tb3+ organic framework series for marking in gunshot residues. J Phys Chem C. 2022;126(38):16568–77.
85. Harshey A, Das T, Srivastava A. Analytical contributions of lanthanide based metal-organic frame works as luminescent markers: recent trends in gunshot residue analysis. Microchem J. 2020;154:104597.

The manufacturer's authorised representative in the EU is Springer Nature Customer Service Centre GmbH, Europaplatz 3, 69115 Heidelberg, Germany. If you have any concerns regarding our products, please contact ProductSafety@springernature.com

Printed and bound by CPI Group (UK) Ltd, Croydon, CR0 4YY
30/09/2026
02225302-0001